# VOICES FROM CHERNOBYL

# VOICES FROM

# CHERNOBYL

## SVETLANA ALEXIEVICH

### TRANSLATION AND PREFACE BY KEITH GESSEN

DALKEY ARCHIVE PRESS

MCLEAN — DUBLIN

First published in Russian as *Tchernobylskaia Molitva* by Editions Ostojie, 1997
Copyright © 1997 by Svetlana Alexievich
Translation © 2005 by Keith Gessen

First edition, 2005

First paperback edition, 2019


Library of Congress Cataloging-in-Publication Data available

ISBN-13 978-1-62897-330-3

Partially funded by grants from the Lannan Foundation, the National Endowment
for the Arts, a federal agency, and the Illinois Arts Council, a state agency.

**www.dalkeyarchive.com**

McLean, IL / Dublin

# CONTENTS

## PART THREE: AMAZED BY SADNESS

# TRANSLATOR'S PREFACE

On September 11, 2001, after the first hijacked plane hit the World Trade Center, emergency triage stations were set up throughout New York City. Doctors and nurses rushed to their hospitals for extra shifts, and many individuals came to donate blood. These were touching acts of generosity and solidarity. The shocking thing about them was that the blood and triage stations turned out to be unnecessary. There were few survivors of the collapse of the two towers.

The effects of the explosion and nuclear fire at the Chernobyl power plant in 1986 were the exact opposite. The initial blast killed just one plant worker, Valeriy Khodomchuk, and in the next few weeks fewer than thirty workers and firemen died from acute radiation poisoning. But tens of thousands received extremely high doses of radiation—it was an accident that produced, in a way, more survivors than victims—and this book is about them.

Much of the material collected here is obscene. In the very first interview, Lyudmilla Ignatenko, the wife of a fireman whose brigade was the first to arrive at the reactor, talks about the total degeneration of her husband's very skin in the week before his death, describing a process so unnatural we should never have had to witness it. "Any little knot [in his bedding], that was already a wound on him," she says. "I clipped my nails down till they bled so I wouldn't accidentally cut him."

Some of the interviews are macabre. Viktor Iosifovich Verzhikovskiy, head of the Khoyniki Society of Volunteer Hunters and Fisherman, recalls his meeting with the regional Party bosses a few months after the explosion. They explained that the Zone of Exclusion, as the Soviets termed the land within thirty kilometers of the Chernobyl power plant, evacuated of humans, was still filled with household pets. But the dogs and cats had absorbed heavy doses of radiation in their fur, and were liable, presumably, to wander out of the Zone. The hunters had to go in and shoot them all. Several other accounts, particularly those about the "deactivation" of the physical landscape in the Zone—the digging up of earth and trees and houses and their (haphazard) burial as nuclear waste—also have this quasi-Gogolian sense: they are ordinary human activities gone terribly berserk.

But in the end it's the very quotidian ordinariness of these testimonies that makes them such a unique human document. "I know you're curious," says Arkady Filin, impressed into Chernobyl service as a "liquidator," or clean-up crew member. "People who weren't there are always curious. But it was still a world of people. The men drank vodka. They played cards, tried to get girls." Or, in the words of one of the hunters: "If you ran over a turtle with your jeep, the shell held up. It didn't crack. Of course we only did this when we were drunk." Even the most desperate cases are still very much part of this "world of people," with its people problems and people worries. "When I die," Valentina Timofeevna Panasevich's husband, also a "liquidator," tells her as he succumbs to cancer several years after his stint at Chernobyl, "sell the car and the spare tire. And don't marry Tolik." Tolik is his brother. Valentina does not marry him.

*

Svetlana Alexievich collected these interviews in 1996—a time when anti-Communism still had some currency as a political idea in the post-Soviet space. And it's certainly true that Chernobyl, while an accident in the sense that no one intentionally set it off, was also the deliberate product of a culture of cronyism, laziness, and a deep-seated indifference toward the general population. The literature on the subject is pretty unanimous in its opinion that the Soviet system had taken a poorly designed reactor and then staffed it with a group of incompetents. It then proceeded, as the interviews in this book attest, to lie about the disaster in the most criminal way. In the crucial first ten days, when the reactor core was burning and releasing a steady stream of highly radioactive material into the surrounding area, the authorities repeatedly claimed that the situation was under control. "If I'd known he'd get sick I'd have closed all the doors," one of the Chernobyl war widows tells Alexievich about her husband, who went to Chernobyl as a liquidator. "I'd have stood in the doorway. I'd have locked the doors with all the locks we had." But no one knew.

And yet, as these testimonies also make all too clear, it wasn't as if the Soviets simply let Chernobyl burn. This is the remarkable thing. On the one hand, total incompetence, indifference, and out-and-out lies. On the other, a genuinely frantic effort to deal with the consequences. In the week after the accident, while refusing to admit to the world that anything really serious had gone wrong, the Soviets poured thousands of men into the breach. They dropped bags of

sand onto the reactor fire from the open doors of helicopters
(analysts now think this did more harm than good). When
the fire stopped, they climbed onto the roof and cleared the
radioactive debris. The machines they brought broke down
because of the radiation. The humans wouldn't break down
until weeks or months later, at which point they'd die horribly.
In 1986 the Soviets threw untrained and unprotected men at
the reactor just as in 1941 they'd thrown untrained, unarmed
men at the *Wehrmacht*, hoping the Germans would at least
have to stop long enough to shoot them. But as the curator of
the Chernobyl Museum correctly explains, had this effort not
been made, the catastrophe might have been a lot worse.

*

In Belarus, very little has changed since these interviews were
conducted. Back in 1996, Aleksandr Lukashenka was the
lesser-known of Europe's "last two dictators." Now Slobodan
Milosevic is on trial at The Hague and Lukashenka has pride
of place. He stifles any attempt at free speech and his political
opponents continue to "disappear." On the Chernobyl front,
Lukashenka has encouraged studies arguing that the land
is increasingly safe and that more and more of it should be
brought back into agricultural rotation. In 1999, the physicist
Yuri Bandazhevsky, a friend and colleague of Vasily Borisovich
Nesterenko (interviewed on page 210), authored a report criti-
cizing this tendency in government policy and suggesting that
Belarus was knowingly exporting contaminated food. He has
been in jail ever since.

*—Keith Gessen, 2005*

# VOICES FROM CHERNOBYL

# HISTORICAL NOTE

There are no nuclear power stations in Belarus. Of the functioning stations in the territory of the former USSR, the ones closest to Belarus are of the old Soviet-designed RBMK type. To the north, the Ignalinsk station, to the east, the Smolensk station, and to the south, Chernobyl.

On April 26, 1986, at 1:23:58, a series of explosions destroyed the reactor in the building that housed Energy Block #4 of the Chernobyl Nuclear Power Station. The catastrophe at Chernobyl became the largest technological disaster of the twentieth century.

For tiny Belarus (population: 10 million), it was a national disaster. During the Second World War, the Nazis destroyed 619 Belarussian villages along with their inhabitants. As a result of Chernobyl, the country lost 485 villages and settlements. Of these, 70 have been forever buried underground. During the war, one out of every four Belarussians was killed; today, one out of every five Belarussians lives on contaminated land. This amounts to 2.1 million people, of whom 700,000 are children. Among the demographic factors responsible for the depopulation of Belarus, radiation is number one. In the Gomel and Mogilev regions, which suffered the most from Chernobyl, mortality rates exceed birth rates by 20%.

As a result of the accident, 50 million Ci of radionuclides were released into the atmosphere. Seventy percent of these descended on Belarus; fully 23% of its territory is contaminated by cesium-137 radionuclides with a density of over 1 Ci/km². Ukraine on the other hand has 4.8% of its territory contaminated, and Russia, 0.5%. The area of arable land with a density of more than 1 Ci/km² is over 18 million hectares; 2.4 thousand hectares have been taken out of the agricultural economy. Belarus is a land of forests. But 26% of all forests and a large part of all marshes near the rivers Pripyat, Dniepr, and Sozh are considered part of the radioactive zone. As a result of the perpetual presence of small doses of radiation, the number of people with cancer, mental retardation, neurological disorders, and genetic mutations increases with each year.

—"Chernobyl." *Belaruskaya entsiklopedia*

On April 29, 1986, instruments recorded high levels of radiation in Poland, Germany, Austria, and Romania. On April 30, in Switzerland and northern Italy. On May 1 and 2, in France, Belgium, the Netherlands, Great Britain, and northern Greece. On May 3, in Israel, Kuwait, and Turkey. . . . Gaseous airborne particles traveled around the globe: on May 2 they were registered in Japan, on May 5 in India, on May 5 and 6 in the U.S. and Canada. It took less than a week for Chernobyl to become a problem for the entire world.

—"The Consequences of the Chernobyl Accident in Belarus." Minsk, Sakharov International College on Radioecology

The fourth reactor, now known as the Cover, still holds about twenty tons of nuclear fuel in its lead-and-metal core. No one knows what is happening with it.

The sarcophagus was well made, uniquely constructed, and the design engineers from St. Petersburg should probably be proud. But it was constructed in absentia, the plates were put together with the aid of robots and helicopters, and as a result there are fissures. According to some figures, there are now over 200 square meters of spaces and cracks, and radioactive particles continue to escape through them . . .

Might the sarcophagus collapse? No one can answer that question, since it's still impossible to reach many of the connections and constructions in order to see if they're sturdy. But everyone knows that if the Cover were to collapse, the consequences would be even more dire than they were in 1986.

—*Ogonyok* magazine, No. 17, April 1996

# A SOLITARY HUMAN VOICE

We are air, we are not earth . . .
—M. Mamardashvili

I don't know what I should talk about—about death or about love? Or are they the same? Which one should I talk about?

We were newlyweds. We still walked around holding hands, even if we were just going to the store. I would say to him, "I love you." But I didn't know then how much. I had no idea . . . We lived in the dormitory of the fire station where he worked. On the second floor. There were three other young couples, we all shared a kitchen. On the first floor they kept the trucks. The red fire trucks. That was his job. I always knew what was happening—where he was, how he was.

One night I heard a noise. I looked out the window. He saw me. "Close the window and go back to sleep. There's a fire at the reactor. I'll be back soon."

I didn't see the explosion itself. Just the flames. Everything was radiant. The whole sky. A tall flame. And smoke. The heat was awful. And he's still not back.

The smoke was from the burning bitumen, which had covered the roof. He said later it was like walking on tar. They tried to beat down the flames. They kicked at the burning graphite with their feet. . . . They weren't wearing their canvas gear. They went off just as they were, in their shirt sleeves. No one told them. They had been called for a fire, that was it.

Four o'clock. Five. Six. At six we were supposed to go to his parents' house. To plant potatoes. It's forty kilometers from Pripyat to Sperizhye, where his parents live. Sowing, plowing—he loved to do that. His mother always told me how they didn't want him to move to the city, they'd even built a new house for him. He was drafted into the army. He served in the fire brigade in Moscow and when he came out, he wanted to be a fireman. And nothing else! [*Silence.*]

Sometimes it's as though I hear his voice. Alive. Even photographs don't have the same effect on me as that voice. But he never calls to me . . . not even in my dreams. I'm the one who calls to him.

Seven o'clock. At seven I was told he was in the hospital. I ran there, but the police had already encircled it, and they weren't letting anyone through. Only ambulances. The policemen shouted: the ambulances are radioactive, stay away! I wasn't the only one there, all the wives whose husbands were at the reactor that night had come. I started looking for a friend, she was a doctor at that hospital. I grabbed her white coat when she came out of an ambulance. "Get me inside!" "I can't. He's bad. They all are." I held on to her. "Just to see him!" "All right," she said. "Come with me. Just for fifteen or twenty minutes."

I saw him. He was all swollen and puffed up. You could barely see his eyes.

"He needs milk. Lots of milk," my friend said. "They should drink at least three liters each." "But he doesn't like milk." "He'll drink it now." Many of the doctors and nurses in that hospital, and especially the orderlies, would get sick themselves and die. But we didn't know that then.

At ten in the morning, the cameraman Shishenok died. He was the first. On the first day. We learned that another one was left under the debris—Valera Khodemchuk. They never did reach him. They buried him under the concrete. And we didn't know then that they were just the first ones.

I said, "Vasya, what should I do?" "Get out of here! Go! You have our child." But how can I leave him? He's telling me: "Go! Leave! Save the baby." "First I need to bring you some milk, then we'll decide what to do." My friend Tanya Kibenok comes running in—her husband's in the same room. Her father's with her, he has a car. We get in and drive to the nearest village for some milk. It's about three kilometers from the town. We buy a bunch of three-liter bottles, six, so it's enough for everyone. But they started throwing up from the milk. They kept passing out, they got put on IVs. The doctors kept telling them they'd been poisoned by gas. No one said anything about radiation. And the town was inundated right away with military vehicles, they closed off all the roads. The trolleys stopped running, and the trains. They were washing the streets with some white powder. I worried about how I'd get to the village the next day to buy some more fresh milk. No one talked about the radiation. Only the military people wore surgical masks. The people in town were carrying bread from the stores, just open sacks with the loaves in them. People were eating cupcakes on plates.

I couldn't get into the hospital that evening. There was a sea of people. I stood under his window, he came over and yelled something to me. It was so desperate! Someone in the crowd heard him—they were being taken to Moscow that night. All the wives got together in one group. We decided we'd go with them. Let us go with our husbands! You have no right! We punched and clawed. The soldiers—there were already soldiers—they pushed us back. Then the doctor came out and said, yes, they were flying to Moscow, but we needed to bring them their clothes. The clothes they'd worn at the station had been burned. The buses had stopped running already and we ran across the city. We came running back with their bags, but the plane was already gone. They tricked us. So that we wouldn't be there yelling and crying.

It's night. On one side of the street there are buses, hundreds of buses, they're already preparing the town for evacuation, and on the other side, hundreds of fire trucks. They came from all over. And the whole street covered in white foam. We're walking on it, just cursing and crying. Over the radio they tell us they might evacuate the city for three to five days, take your warm clothes with you, you'll be living in the forest. In tents. People were even glad—a camping trip! We'll celebrate May Day like that, a break from routine. People got barbeques ready. They took their guitars with them, their radios. Only the women whose husbands had been at the reactor were crying.

I can't remember the trip out to my parents' village. It was like I woke up when I saw my mother. "Mama. Vasya's in Moscow. They flew him out on a special plane!" But we finished planting the garden. [*A week later the village was evacuated.*] Who knew? Who knew that then? Later in the day I started throwing up. I was six months pregnant. I felt awful. That night I dreamed he was calling out to me in his sleep: "Lyusya! Lyusenka!" But after he died, he didn't call out in my dreams anymore. Not once. [*She starts crying.*] I got up in the morning thinking I have to get to Moscow. By myself. My mother's crying: "Where are you going, the way you are?" So I took my father with me. He went to the bank and took out all the money they had.

I can't remember the trip. The trip just isn't in my memory. In Moscow we asked the first police officer we saw, Where did they put the Chernobyl firemen, and he told us. We were surprised, too, everyone was scaring us that it was top secret. "Hospital number 6. At the Shchukinskaya stop."

It was a special hospital, for radiology, and you couldn't get in without a pass. I gave some money to the woman at the door, and she said, "Go ahead." Then I had to ask someone else, beg. Finally I'm sitting in the office of the head radiologist, Angelina Vasilyevna Guskova. But I didn't know that yet, what her name was, I didn't remember anything. I just knew I had to see him. Right away she asked: "Do you have kids?"

What should I tell her? I can see already I need to hide that I'm pregnant. They won't let me see him! It's good I'm thin, you can't really tell anything.

"Yes," I say.

"How many?"

I'm thinking, "I need to tell her two. If it's just one, she won't let me in."

"A boy and a girl."

"So you don't need to have anymore. All right, listen: his central nervous system is completely compromised, his skull is completely compromised."

Okay, I'm thinking, so he'll be a little fidgety.

"And listen: if you start crying, I'll kick you out right away. No hugging or kissing. Don't even get near him. You have half an hour."

But I knew already that I wasn't leaving. If I leave, then it'll be with him. I swore to myself! I come in, they're sitting on the bed, playing cards and laughing.

"Vasya!" they call out.

He turns around:

"Oh, well, now it's over! Even here she found me!"

He looks so funny, he's got pajamas on for a size 48, and he's a size 52. The sleeves are too short, the pants are too short. But his face isn't swollen anymore. They were given some sort of fluid.

I say, "Where'd you run off to?"

He wants to hug me.

The doctor won't let him. "Sit, sit," she says. "No hugging in here."

We turned it into a joke somehow. And then everyone comes over, from the other rooms too, everyone from Pripyat. There were twenty-eight of them on the plane. What's going on? How are things in town? I tell them they've begun evacuating everyone, the whole town is being cleared out for three or five days. None of the guys says anything, and then one of the women, there were two women, she was on duty at the factory the day of the accident, she starts crying.

"Oh God! My kids are there. What's happening with them?"

I wanted to be with him alone, if only for a minute. The guys felt it, and each of them thought of some excuse, and they all went out into the hall. Then I hugged and kissed him. He moved away.

"Don't sit near me. Get a chair."

"That's just silly," I said, waving it away. "Did you see the explosion? Did you see what happened? You were the first ones there."

"It was probably sabotage. Someone set it up. All the guys think so."

That's what people were saying then. That's what they thought.

The next day, they were lying by themselves, each in his own room. They were banned from going in the hallway, from talking to each other. They knocked on the walls with their knuckles. Dash-dot, dash-dot. The doctors explained that everyone's body reacts differently to radiation, and one person can handle what another can't. They even measured the radiation of the walls where they had them. To the right, left, and the floor beneath. They moved out all the sick people from the floor below and the floor above. There was no one left in the place.

For three days I lived with my friends in Moscow. They kept saying: Take the pot, take the plate, take whatever you need. I made turkey soup for six. For six of our boys. Firemen. From the same shift. They were all on duty that night: Bashuk, Kibenok, Titenok, Pravik, Tischura. I went to the store and bought them toothpaste and toothbrushes and soap. They didn't have any of that at the hospital. I bought them little towels. Looking back, I'm surprised by my friends: they were afraid, of course, how could they not be, there were rumors already, but still they kept saying: Take whatever you need, take it! How is he? How are they all? Will they live? Live. [*She is silent.*] I met a lot of good people then, I don't remember all of them. I remember an old woman janitor, who taught me: "There are sicknesses that can't be cured. You just have to sit and watch them."

Early in the morning I go to the market, then to my friends' place, where I make the soup. I have to grate everything and grind it. Someone said, "Bring me some apple juice." So I come with six half-liter cans, always for six! I race to the hospital, then I sit there until evening. In the evening, I go back across the city. How much longer could I have kept that up? After three days they told me I could stay in the dorm for medical workers, it's on hospital grounds. God, how wonderful!

"But there's no kitchen. How am I going to cook?"

"You don't need to cook anymore. They can't digest the food."

He started to change—every day I met a brand-new person. The burns started to come to the surface. In his mouth, on his tongue, his cheeks—at first there were little lesions, and then they grew. It came off in layers—as white film . . . the color of his face . . . his body . . . blue . . . red . . . gray-brown. And it's all so very mine! It's impossible to describe! It's impossible to write down! And even to get over. The only thing that saved me was it happened so fast; there wasn't any time to think, there wasn't any time to cry.

I loved him! I had no idea how much! We'd just gotten married. We're walking down the street—he'd grab my hands and whirl me around. And kiss me, kiss me. People are walking by and smiling.

It was a hospital for people with serious radiation poisoning. Fourteen days. In fourteen days a person dies.

On the very first day in the dormitory they measured me with a dosimeter. My clothes, bag, purse, shoes—they were all "hot." And they took that all away from me right there. Even my underthings. The only thing they left was my money. In exchange they gave me a hospital robe—a size 56—and some size 43 slippers. They said they'd return the clothes, maybe, or maybe they wouldn't, since they might not be possible to "launder" at this point. That is how I looked when I came to visit him. I frightened him. "Woman, what's wrong with you?" But I was still able to make him some soup. I boiled the water in a glass jar, and then I threw pieces of chicken in there—tiny, tiny pieces. Then someone gave me her pot, I think it was the cleaning woman or the guard. Someone else gave me a cutting board, for chopping my parsley. I couldn't go to the market in my hospital robe, people would bring me the vegetables. But it was all useless, he couldn't even drink anything. He couldn't even swallow a raw egg. But I wanted to get something tasty! As if it mattered. I ran to the post office. "Girls," I told them, "I need to call my parents in Ivano-Frankovsk right away! My husband is dying." They understood right away where I was from and who my husband was, and they connected me. My father, sister, and brother flew out that very day to Moscow. They brought me my things. And money. It was the ninth of May. He always used to say to me: "You have no idea how beautiful Moscow is! Especially on V-Day, when they set off the fireworks. I want you to see it."

I'm sitting with him in the room, he opens his eyes. "Is it day or night?"

"It's nine at night."

"Open the window! They're going to set off the fireworks!"

I opened the window. We're on the eighth floor, and the whole city's there before us! There was a bouquet of fire exploding in the air.

"Look at that!" I said.

"I told you I'd show you Moscow. And I told you I'd always give you flowers on holidays . . ."

I look over, and he's getting three carnations from under his pillow. He gave the nurse money, and she bought them.

I run over to him and I kiss him.

"My love! My one and only!"

He starts growling. "What did the doctors tell you? No hugging me. And no kissing!"

They wouldn't let me hug him. But I . . . I lifted him and sat him up. I made his bed. I placed the thermometer. I picked up and brought back the sanitation dish. I stayed up with him all night.

It's a good thing that it was in the hallway, not the room, that my head started spinning, I grabbed onto the windowsill. A doctor was walking by, he took me by the arm. And then suddenly: "Are you pregnant?"

"No, no!" I was so scared someone would hear us.

"Don't lie," he sighed.

The next day I get called to the head doctor's office.

"Why did you lie to me?" she says.

"There was no other way. If I'd told you, you'd send me home. It was a sacred lie!"

"What have you done?"

"But I was with him . . ."

I'll be grateful to Angelina Vasilyevna Guskova my whole life. My whole life! Other wives also came, but they weren't allowed in. Their mothers were with me. Volodya Pravik's mother kept begging God: "Take me instead." An American professor, Dr. Gale—he's the one who did the bone marrow operation—tried to comfort me. There's a tiny ray of hope, he said, not much, but a little. Such a powerful organism, such a strong guy! They called for all his relatives. Two of his sisters came from Belarus, his brother from Leningrad, he was in the army there. The younger one, Natasha, she was fourteen, she was very scared and cried a lot. But her bone marrow was the best fit. [Silent.] Now I can talk about this. Before I couldn't. I didn't talk about it for ten years. [Silent.]

When he found out they'd be taking the bone marrow from his little sister, he flat-out refused. "I'd rather die. She's so small. Don't touch her." His older sister Lyuda was twenty-eight, she was a nurse herself, she knew what she was getting into. "As long as he lives," she said. I watched the operation. They were lying next to each other on the tables. There was a big window onto the operating room. It took two hours. When they were done, Lyuda was worse off than he was, she had eighteen punctures in her chest, it was very difficult for her to come out from under the anesthesia. Now she's sick, she's an invalid. She was a strong, pretty girl. She never got married. So then I was running from one room to the other, from his room to hers. He wasn't in an ordinary room anymore, he was in a special bio-chamber, behind a transparent curtain. No one was allowed inside.

They have instruments there, so that without going through the curtain they can give him shots, place the catheter. The curtains are held together by Velcro, and I've learned to use them. But I push them aside and go inside to him. There was a little chair next to his bed. He got so bad that I couldn't leave him now even for a second. He was calling me constantly: "Lyusya, where are you? Lyusya!" He called and called. The other bio-chambers, where our boys were, were being tended to by soldiers, because the orderlies on staff refused, they demanded protective clothing. The soldiers carried the sanitary vessels. They wiped the floors down, changed the bedding. They did everything. Where did they get those soldiers? We didn't ask. But he—he—every day I would hear: Dead. Dead. Tischura is dead. Titenok is dead. Dead. It was like a sledgehammer to my brain.

He was producing stool 25 to 30 times a day. With blood and mucous. His skin started cracking on his arms and legs. He became covered with boils. When he turned his head, there'd be a clump of hair left on the pillow. I tried joking: "It's convenient, you don't need a comb." Soon they cut all their hair. I did it for him myself. I wanted to do everything for him myself. If it had been physically possible I would have stayed with him all twenty-four hours. I couldn't spare a minute. [*Long silence.*] My brother came and he got scared. "I won't let you in there!" But my father said to him: "You think you can stop her? She'll go through the window! She'll get up through the fire escape!"

I go back to the hospital and there's an orange on the bedside table. A big one, and pink. He's smiling: "I got a gift. Take it." Meanwhile the nurse is gesturing through the film that I can't eat it. It's been near him a while, so not only can you not eat it, you shouldn't even touch it. "Come on, eat it," he says. "You like oranges." I take the orange in my hand. Meanwhile he shuts his eyes and goes to sleep. They were always giving him shots to put him to sleep. The nurse is looking at me in horror. And me? I'm ready to do whatever it takes so that he doesn't think about death. And about the fact that his death is horrible, that I'm afraid of him. There's a fragment of some conversation, I'm remembering it. Someone is saying: "You have to understand: this is not your husband anymore, not a beloved person, but a radioactive object with a strong density of poisoning. You're not suicidal. Get ahold of yourself." And I'm like someone who's lost her mind: "But I love him! I love him!" He's sleeping, and I'm whispering: "I love you!" Walking in the hospital courtyard, "I love you." Carrying his sanitary tray, "I love you." I remembered how we used to live at home. He only fell asleep at night after he'd taken my hand. That was a habit of his—to hold my hand while he slept. All night. So in the hospital I take his hand and don't let go.

One night, everything's quiet. We're all alone. He looked at me very, very carefully and suddenly he said:

"I want to see our child so much. How is he?"

"What are we going to name him?"

"You'll decide that yourself."

"Why myself, when there's two of us?"

"In that case, if it's a boy, he should be Vasya, and if it's a girl, Natasha."

I had no idea then how much I loved him! Him . . . just him. I was like a blind person! I couldn't even feel the little pounding underneath my heart. Even though I was six months in. I thought that my little one was inside me, that he was protected.

None of the doctors knew I was staying with him at night in the bio-chamber. The nurses let me in. At first they pleaded with me, too: "You're young. Why are you doing this? That's not a person anymore, that's a nuclear reactor. You'll just burn together." I was like a dog, running after them. I'd stand for hours at their doors, begging and pleading. And then they'd say: "All right! The hell with you! You're not normal!" In the mornings, just before eight, when the doctors started their rounds, they'd be there on the other side of the film: "Run!" So I'd go to the dorm for an hour. Then from 9 A.M. to 9 P.M. I have a pass to come in. My legs were blue below the knee, blue and swollen, that's how tired I was.

While I was there with him, they wouldn't, but when I left—they photographed him. Without any clothes. Naked. One thin little sheet on top of him. I changed that little sheet every day, and every day by evening it was covered in blood. I pick him up, and there are pieces of his skin on my hand, they stick to my hands. I ask him: "Love. Help me. Prop yourself up on your arm, your elbow, as much as you can, I'll smooth out your bedding, get the knots and folds out." Any little knot, that was already a wound on him. I clipped my nails down till they bled so I wouldn't accidentally cut him. None of the nurses could approach him; if they needed anything they'd call me.

And they photographed him. For science, they said. I'd have pushed them all out of there! I'd have yelled! And hit them! How dare they? It's all mine—it's my love—if only I'd been able to keep them out of there.

I'm walking out of the room into the hallway. And I'm walking toward the couch, because I don't see them. I tell the nurse on duty: "He's dying." And she says to me: "What did you expect? He got 1,600 roentgen. Four hundred is a lethal dose. You're sitting next to a nuclear reactor." It's all mine . . . it's my love. When they all died, they did a *remont* at the hospital. They scraped down the walls and dug up the parquet.

And then—the last thing. I remember it in flashes, all broken up.

I'm sitting on my little chair next to him at night. At eight I say: "Vasenka, I'm going to go for a little walk." He opens his eyes and closes them, lets me go. I just walk to the dorm, go up to my room, lie down on the floor, I couldn't lie on the bed, everything hurt too much, when already the cleaning lady is knocking on the door. "Go! Run to him! He's calling for you like mad!" That morning Tanya Kibenok pleaded with me: "Come to the cemetery, I can't go there alone." They were burying Vitya Kibenok and Volodya Pravik. They were friends of my Vasya. Our families were friends. There's a photo of us all in the building the day before the explosion. Our husbands are so handsome! And happy! It was the last day of that life. We were all so happy!

I came back from the cemetery and called the nurse's post right away. "How is he?" "He died fifteen minutes ago." What? I was there all night. I was gone for three hours! I came up to the window and started shouting: "Why? Why?" I looked up at the sky and yelled. The whole building could hear me. They were afraid to come up to me. Then I came to: I'll see him one more time! Once more! I run down the stairs. He was still in his bio-chamber, they hadn't taken him away yet. His last words were "Lyusya! Lyusenka!" "She's just stepped away for a bit, she'll be right back," the nurse told him. He sighed and went quiet. I didn't leave him anymore after that. I escorted him all the way to the grave site. Although the thing I remember isn't the grave, it's the plastic bag. That bag.

At the morgue they said, "Want to see what we'll dress him in?" I do! They dressed him up in formal wear, with his service cap. They couldn't get shoes on him because his feet had swelled up. They had to cut up the formal wear, too, because they couldn't get it on him, there wasn't a whole body to put it on. It was all—wounds. The last two days in the hospital—I'd lift his arm, and meanwhile the bone is shaking, just sort of dangling, the body has gone away from it. Pieces of his lungs, of his liver, were coming out of his mouth. He was choking on his internal organs. I'd wrap my hand in a bandage and put it in his mouth, take out all that stuff. It's impossible to talk about. It's impossible to write about. And even to live through. It was all mine. My love. They couldn't get a single pair of shoes to fit him. They buried him barefoot.

Right before my eyes—in his formal wear—they just took him and put him in that cellophane bag of theirs and tied it up. And then they put this bag in the wooden coffin. And they tied the coffin with another bag. The plastic is transparent, but thick, like a tablecloth. And then they put all that into a zinc coffin. They squeezed it in. Only the cap didn't fit.

Everyone came—his parents, my parents. They bought black handkerchiefs in Moscow. The Extraordinary Commission met with us. They told everyone the same thing: it's impossible for us to give you the bodies of your husbands, your sons, they are very radioactive and will be buried in a Moscow cemetery in a special way. In sealed zinc caskets, under cement tiles. And you need to sign this document here.

If anyone got indignant and wanted to take the coffin back home, they were told that the dead were now, you know, heroes, and that they no longer belonged to their families. They were heroes of the State. They belonged to the State.

We sat in the hearse. The relatives and some sort of military people. A colonel and his regiment. They tell the regiment: "Await your orders!" We drive around Moscow for two or three hours, around the beltway. We're going back to Moscow again. They tell the regiment: "We're not allowing anyone into the cemetery. The cemetery's being attacked by foreign correspondents. Wait some more." The parents don't say anything. Mom has a black handkerchief. I sense I'm about to black out. "Why are they hiding my husband? He's—what? A murderer? A criminal? Who are we burying?" My mom: "Quiet. Quiet, daughter." She's petting me on the head. The colonel calls in: "Let's enter the cemetery. The wife is getting hysterical." At the cemetery we were surrounded by soldiers. We had a convoy. And they were carrying the coffin. No one was allowed in. It was just us. They covered him with earth in a minute. "Faster! Faster!" the officer was yelling. They didn't even let me hug the coffin. And—onto the bus. Everything on the sly.

Right away they bought us plane tickets back home. For the next day. The whole time there was someone with us in plainclothes with a military bearing. He wouldn't even let us out of the dorm to buy some food for the trip. God forbid we might talk with someone—especially me. As if I could talk by then. I couldn't even cry. When we were leaving, the woman on duty counted all the towels and all the sheets. She folded them right away and put them into a polyethylene bag. They probably burnt them. We paid for the dormitory ourselves. For fourteen nights. It was a hospital for radiation poisoning. Fourteen nights. That's how long it takes a person to die.

At home I fell asleep. I walked into the place and just fell onto the bed. I slept for three days. An ambulance came. "No," said the doctor, "she'll wake up. It's just a terrible sleep."

I was twenty-three.

I remember the dream I had. My dead grandmother comes to me in the clothes that we buried her in. She's dressing up the New Year's tree. "Grandma, why do we have a New Year's tree? It's summertime." "Because your Vasenka is going to join me soon." And he grew up in the forest. I remember the dream—Vasya comes in a white robe and calls for Natasha. That's our girl, who I haven't given birth to yet. She's already grown up. He throws her up to the ceiling, and they laugh. And I'm watching them and thinking that happiness—it's so simple. I'm sleeping. We're walking along the water. Walking and walking. He probably asked me not to cry. Gave me a sign. From up there.

[*She is silent for a long time.*]

Two months later I went to Moscow. From the train station straight to the cemetery. To him! And at the cemetery I start going into labor. Just as I started talking to him—they called the ambulance. It was at the same Angelina Vasilyevna Guskova's that I gave birth. She'd said to me back then: "You need to come here to give birth." It was two weeks before I was due.

They showed her to me—a girl. "Natashenka," I called out. "Your father named you Natashenka." She looked healthy. Arms, legs. But she had cirrhosis of the liver. Her liver had twenty-eight roentgen. Congenital heart disease. Four hours later they told me she was dead. And again: we won't give her to you. What do you mean you won't give her to me? It's me who won't give her to you! You want to take her for science. I hate your science! I hate it!

[*She is silent.*]

I keep saying the wrong thing to you. The wrong thing. I'm not supposed to yell after my stroke. And I'm not supposed to cry. That's why the words are all wrong. But I'll say this. No one knows this. When they brought me the little wooden box and said, "She's in there," I looked. She'd been cremated. She was ashes. And I started crying. "Put her at his feet," I requested.

There, at the cemetery, it doesn't say Natasha Ignatenko. There's only his name. She didn't have a name yet, she didn't have anything. Just a soul. That's what I buried there. I always go there with two bouquets: one for him, and the other I put in the corner for her. I crawl around the grave on my knees. Always on my knees. [*She becomes incomprehensible.*] I killed her. I. She. Saved. My little girl saved me, she took the whole radioactive shock into herself, she was like the lightning rod for it. She was so small. She was a little tiny thing. [*She has trouble breathing.*] She saved . . . But I loved them both. Because—because you can't kill something with love, right? With such love! Why are these things together—love and death. Together. Who's going to explain this to me? I crawl around the grave on my knees.

[*She is silent for a long time.*]

In Kiev they gave me an apartment. It was in a large building, where they put everyone from the atomic station. It's a big apartment, with two rooms, the kind Vasya and I had dreamed of. And I was going crazy in it!

I found a husband eventually. I told him everything—the whole truth—that I have one love, for my whole life. I told him everything. We'd meet, but I'd never invite him to my home, that's where Vasya was.

I worked in a candy shop. I'd be making cake, and tears would be rolling down my cheeks. I'm not crying, but there are tears rolling down.

I gave birth to a boy, Andrei. Andreika. My friends tried to stop me. "You can't have a baby." And the doctors tried to scare me: "Your body won't be able to handle it." Then, later—later they told me that he'd be missing an arm. His right arm. The instrument showed it. "Well, so what?" I thought. "I'll teach him to write with his left hand." But he came out fine. A beautiful boy. He's in school now, he gets good grades. Now I have someone—I can live and breathe him. He's the light in my life. He understands everything perfectly. "Mom, if I go visit grandma for two days, will you be able to breathe?" I won't! I fear the day I'll have to leave him. One day we're walking down the street. And I feel that I'm falling. That's when I had my first stroke. Right on the street. "Mom, do you need some water?" "No, just stand here next to me. Don't go anywhere." And I grabbed his arm. I don't remember what happened next. I came to in the hospital. But I grabbed him so hard that the doctors were barely able to pry my fingers open. His arm was blue for a long time. Now we walk out of the house, he says, "Mommie, just don't grab my arm. I won't go anywhere." He's also sick: two weeks in school, two weeks at home with a doctor. That's how we live.

[*She stands up, goes over to the window.*]

There are many of us here. A whole street. That's what it's called—Chernobylskaya. These people worked at the station their whole lives. A lot of them still go there to work on a provisional basis, that's how they work there now, no one lives there anymore. They have bad diseases, they're invalids, but they don't leave their jobs, they're scared to even think of the reactor closing down. Who needs them now anywhere else? Often they die. In an instant. They just drop—someone will be walking, he falls down, goes to sleep, never wakes up. He was carrying flowers for his nurse and his heart stopped. They die, but no one's really asked us. No one's asked what we've been through. What we saw. No one wants to hear about death. About what scares them.

But I was telling you about love. About my love . . .

*Lyudmilla Ignatenko,*
*wife of deceased fireman Vasily Ignatenko*

# PART ONE

# THE LAND OF THE DEAD

## MONOLOGUE ON WHY WE REMEMBER

You've decided to write about this? About *this*? But I wouldn't want people to know this about me, what I went through there. On the one hand, there's the desire to open up, to say everything, and on the other—I feel like I'm exposing myself, and I wouldn't want to do that.

Do you remember how it was in Tolstoy? Pierre Bezukhov is so shocked by the war, he thinks that he and the whole world have changed forever. But then some time passes, and he says to himself: "I'm going to keep yelling at the coach-driver just like before, I'm going to keep growling like before." Then why do people remember? So that they can determine the truth? For fairness? So they can free themselves and forget? Is it because they understand they're part of a grand event? Or are they looking into the past for cover? And all this despite the fact that memories are very fragile things, ephemeral things, this is not exact knowledge, but a guess that a person makes about himself. It isn't even knowledge, it's more like a set of emotions.

My emotions . . . I struggled, I dug into my memory and I remembered.

The scariest thing for me was during my childhood—that was the war.

I remember how we boys played "mom and dad"—we'd take the clothes off the little ones and put them on top of one another. These were the first kids born after the war, because during the war kids were forgotten. We waited for life to appear. We played "mom and dad." We wanted to see how life would appear. We were eight, ten years old.

I saw a woman trying to kill herself. In the bushes by the river. She had a brick and she was hitting herself in the head with it. She was pregnant from an occupying soldier whom the whole village hated. Also, as a boy, I saw a litter of kittens being born. I helped my mother pull a calf from its mother, I led our pig to meet up with a boar. I remember—I remember how they brought my father's body, he had on a sweater, my mother had knit it herself, and he'd been shot by a machine gun, and bloody pieces of something were coming out of that sweater. He lay on our only bed, there was nowhere else to put him. Later he was buried in front of the house. And the earth wasn't cotton, it was heavy clay. From the beds for beetroot. There were battles going on all around. The street was filled with dead people and horses.

For me, those memories are so personal, I've never spoken of them out loud.

Back then I thought of death just as I did of birth. I had the same feeling when I saw a calf come out of a cow—and the kittens were born—as when I saw that woman with the brick in the bushes killing herself. For some reason these seemed to me to be the same things—birth and death.

I remember from my childhood how a house smells when a boar is being cut up. You've just touched me, and I'm already falling into there, falling—into that nightmare. That terror. I'm flying into it. I also remember how, when we were little, the women would take us with them to the sauna. And we saw that all the women's uteruses (this we could understand even then) were falling out, they were tying them up with rags. I saw this. They were falling out because of hard labor. There were no men, they were at the front, or with the partisans, there were no horses, the women carried all the loads themselves. They ploughed over the gardens themselves, and the kolkhoz fields. When I was older, and I was intimate with a woman, I would remember this—what I saw in the sauna.

I wanted to forget. Forget everything. And I did forget. I thought the most horrible things had already happened. The war. And that I was protected now, that I was protected.

But then I traveled to the Chernobyl Zone. I've been there many times now. And understood how powerless I am. I'm falling apart. My past no longer protects me. There aren't any answers there. They were there before, but now they're not. The future is destroying me, not the past.

*Pyotr S., psychologist*

## MONOLOGUE ABOUT WHAT CAN BE TALKED ABOUT WITH THE LIVING AND THE DEAD

The wolf came into the yard at night. I look out the window and there he is, eyes shining, like headlights. Now I'm used to everything. I've been living alone for seven years, seven years since the people left. Sometimes at night I'll just be sitting here thinking, thinking, until it's lights out again. So on this day I was up all night, sitting on my bed, and then I went out to look at how the sun was. What should I tell you? Death is the fairest thing in the world. No one's ever gotten out of it. The earth takes everyone—the kind, the cruel, the sinners. Aside from that, there's no fairness on earth. I worked hard and honestly my whole life. But I didn't get any fairness. God was dividing things up somewhere, and by the time the line came to me there was nothing left. A young person can die, an old person has to die . . . At first, I waited for people to come—I thought they'd come back. No one said they were leaving forever, they said they were leaving for a while. But now I'm just waiting for death. Dying isn't hard, but it is scary. There's no church. The priest doesn't come. There's no one to tell my sins to.

The first time they told us we had radiation, we thought: it's a sort of a sickness, and whoever gets it dies right away. No, they said, it's this thing that lies on the ground, and gets into the ground, but you can't see it. Animals might be able to see it and hear it, but people can't. But that's not true! I saw it. This cesium was lying in my yard, until it got wet with rain. It was an ink-black color. It was lying there and sort of dripping into pieces. I ran home from the kolkhoz and went into my garden. And there's another piece, it's blue. And 200 meters over, there's another one. About the size of the kerchief on my head. I called over to my neighbor, the other women, we all ran around looking. All the gardens, and the field nearby—about two hectares—we found maybe four big chunks. One was red. The next day it rained early, and by lunchtime they were gone. The police came but there was nothing to show them. We could just tell them. The chunks were like this. [*She indicates the size with her hands.*] Like my kerchief. Blue and red . . .

We weren't too afraid of this radiation. When we couldn't see it, and we didn't know what it was, maybe we were a little afraid, but once we'd seen it, we weren't so afraid. The police and the soldiers put up these signs. Some were next to people's houses, some were in the street—they'd write, 70 curie, 60 curie. We'd always lived off our potatoes, and then suddenly—we're not allowed to! For some people it was real bad, for others it was funny. They advised us to work in our gardens in masks and rubber gloves. And then another big scientist came to the meeting hall and told us that we needed to wash our yards. Come on! I couldn't believe what I was hearing! They ordered us to wash our sheets, our blankets, our curtains. But they're in storage! In closets and trunks. There's no radiation in there! Behind glass? Behind closed doors! Come on! It's in the forest, in the field. They closed the wells, locked them up, wrapped them in cellophane. Said the water was "dirty." How can it be dirty when it's so clean? They told us a bunch of nonsense. You'll die. You need to leave. Evacuate.

People got scared. They got filled up with fear. At night people started packing up their things. I also got my clothes, folded them up. My red badges for my honest labor, and my lucky kopeika that I had. Such sadness! It filled my heart. Let me be struck down right here if I'm lying. And then I hear about how the soldiers were evacuating one village, and this old man and woman stayed. Until then, when people were roused up and put on buses, they'd take their cow and go into the forest. They'd wait there. Like during the war, when they were burning down the villages. Why would our soldiers chase us? [Starts crying.] It's not stable, our life. I don't want to cry.

Oh! Look there—a crow. I don't chase them away. Although sometimes a crow will steal eggs from the barn. I still don't chase them away. I don't chase anyone away! Yesterday a little rabbit came over. There's a village nearby, also there's one woman living there, I said, come by. Maybe it'll help, maybe it won't, but at least there'll be someone to talk to. At night everything hurts. My legs are spinning, like there are little ants running through them, that's my nerve running through me. It's like that when I pick something up. Like wheat being crushed. Crunch, crunch. Then the nerve calms down. I've already worked enough in my life, been sad enough. I've had enough of everything and I don't want anything more.

I have daughters, and sons . . . They're all in the city. But I'm not going anywhere! God gave me years, but he didn't give me a fair share. I know that an old person gets annoying, that the younger generation will run out of patience. I haven't had much joy from my children. The women, the ones who've gone into the city, are always crying. Either their daughter-in-law is hurting their feelings, or their daughter is. They want to come back. My husband is here. He's buried here. If he wasn't lying here, he'd be living in some other place. And I'd be with him. [*Cheers up suddenly.*] And why should I leave? It's nice here! Everything grows, everything blooming. From the littlest fly to the animals, everything's living.

I'll remember everything for you. The planes are flying and flying. Every day. They fly real-real low right over our heads. They're flying to the reactor. To the station. One after the other. While here we have the evacuation. They're moving us out. Storming the houses. People have covered up, they're hiding. The livestock is moaning, the kids are crying. It's war! And the sun's out . . . I sat down and didn't come out of the hut, though it's true I didn't lock up either. The soldiers knocked. "Ma'am, have you packed up?" And I said: "Are you going to tie my hands and feet?" They didn't say anything, didn't say anything, and then they left. They were young. They were kids! Old women were crawling on their knees in front of the houses, begging. The soldiers picked them up under their arms and into the car. But I told them, whoever touched me was going to get it. I cursed at them! I cursed good. I didn't cry. That day I didn't cry. I sat in my house. One minute there's yelling. Yelling! And then it's quiet. Very quiet. On that day—that first day I didn't leave the house.

They told me later that there was a column of people walking. And next to that there was a column of livestock. It was war! My husband liked to say that people shoot, but it's God who delivers the bullet. Everyone has his own fate. The young ones who left, some of them have already died. In their new place. Whereas me, I'm still walking around. Slowing down, sure. Sometimes it's boring, and I cry. The whole village is empty. There's all kinds of birds here. They fly around. And there's elk here, all you want. [*Starts crying.*]

I remember everything. Everyone up and left, but they left their dogs and cats. The first few days I went around pouring milk for all the cats, and I'd give the dogs a piece of bread. They were standing in their yards waiting for their masters. They waited for them a long time. The hungry cats ate cucumbers. They ate tomatoes. Until the fall I took care of my neighbor's lawn, up to the fence. Her fence fell down, I hammered it back up again. I waited for the people. My neighbor had a dog named Zhuchok. "Zhuchok," I'd say, "if you see the people first, give me a shout."

One night I dreamt I was getting evacuated. The officer yells, "Lady! We're going to burn everything down and bury it. Come out!" And they drive me somewhere, to some unknown place. Not clear where. It's not the town, it's not the village. It's not even Earth.

One time—I had a nice little kitty. Vaska. One winter the rats were really hungry and they were attacking. There was nowhere to go. They'd crawl under the covers. I had some grain in a barrel, they put a hole in the barrel. But Vaska saved me. I'd have died without him. We'd talk, me and him, and eat dinner. Then Vaska disappeared. The hungry dogs ate him, maybe, I don't know. They were always running around hungry, until they died. The cats were so hungry they ate their kittens. Not during the summer, but during the winter they would. God, forgive me!

Sometimes now I can't even make it all the way through the house. For an old woman even the stove is cold during the summer. The police come here sometimes, check things out, they bring me bread. But what are they checking for?

It's me and the cat. This is a different cat. When we hear the police, we're happy. We run over. They bring him a bone. Me they'll ask: "What if the bandits come?" "What'll they get off me? What'll they take? My soul? Because that's all I have." They're good boys. They laugh. They brought me some batteries for my radio, now I listen to it. I like Lyudmilla Zykina, but she's not singing as much anymore. Maybe she's old now, like me. My man used to say—he used to say, "The dance is over, put the violin back in the case."

I'll tell you how I found my kitty. I lost my Vaska. I waited a day, two days, then a month. So that was that. I was all alone. No one even to talk to. I walked around the village, going into other people's yards, calling out: Vaska. Murka. Vaska! Murka! At first there were a lot of them running around, and then they disappeared somewhere. Death doesn't care. The earth takes everyone. So I'm walking, and walking. For two days. On the third day I see him under the store. We exchange glances. He's happy, I'm happy. But he doesn't say anything. "All right," I say, "let's go home." But he sits there, meowing. So then I say: "What'll you do here by yourself? The wolves will eat you. They'll tear you apart. Let's go. I have eggs, I have some lard." But how do I explain it to him? Cats don't understand human language, then how come he understood me? I walk ahead, and he runs behind me. Meowing. "I'll cut you off some lard." Meow. "We'll live together the two of us." Meow. "I'll call you Vaska, too." Meow. And we've been living together two winters now.

At night I'll dream that someone's been calling me. The neighbor's voice: "Zina!" Then it's quiet. And again: "Zina!"

I get bored sometimes, and then I cry.

I go to the cemetery. My mom's there. My little daughter. She burned up with typhus during the war. Right after we took her to the cemetery, buried her, the sun came out from the clouds. And shone and shone. Like: you should go and dig her up. My husband is there. Fedya. I sit with them all. I sigh a little. You can talk to the dead just like you can talk to the living. Makes no difference to me. I can hear the one and the other. When you're alone . . . And when you're sad. When you're very sad.

Ivan Prohorovich Gavrilenko, he was a teacher, he lived right next to the cemetery. He moved to the Crimea, his son was there. Next to him was Pyotr Ivanovich Miusskiy. He drove a tractor. He was a Stakhanovite, back then everyone was aching to be a Stakhanovite. He had magic hands. He could make lace out of wood. His house, it was the size of the whole village. Oh, I felt so bad, and my blood boiled, when they tore it down. They buried it. The officer was yelling: "Don't think of it, grandma! It's on a hot-spot!" Meanwhile he's drunk. I come over—Pyotr's crying. "Go on, grandma, it's all right." He told me to go. And the next house is Misha Mikhalev's, he heated the kettles on the farm. He died fast. Left here, and died right away. Next to his house was Stepa Bykhov's, he was a zoologist. It burned down! Bad people burned it down at night. Stepa didn't live long. He's buried somewhere in the Mogilev region. During the war—we lost so many people! Vassily Makarovich Kovalev. Maksim Nikoforenko. They used to live, they were happy. On holidays they'd sing, dance. Play the harmonica. And now, it's like a prison. Sometimes I'll close my eyes and go through the village—well, I say to them, what radiation?

There's a butterfly flying, and bees are buzzing. And my Vaska's catching mice. [*Starts crying.*]

Oh Lyubochka, do you understand what I'm telling you, my sorrow? You'll carry it to people, maybe I won't be here anymore. I'll be in the ground. Under the roots . . .

*Zinaida Yevdokimovna Kovalenko, re-settler*

MONOLOGUE ABOUT A WHOLE LIFE
WRITTEN DOWN ON DOORS

I want to bear witness . . .

It happened ten years ago, and it happens to me again every day.

We lived in the town of Pripyat. In that town.

I'm not a writer. I won't be able to describe it. My mind is not capable of understanding it. And neither is my university degree. There you are: a normal person. A little person. You're just like everyone else—you go to work, you return from work. You get an average salary. Once a year you go on vacation. You're a normal person! And then one day you're suddenly turned into a Chernobyl person. Into an animal, something that everyone's interested in, and that no one knows anything about. You want to be like everyone else, and now you can't. People look at you differently. They ask you: was it scary? How did the station burn? What did you see? And, you know, can you have children? Did your wife leave you? At first we were all turned into animals. The very word "Chernobyl" is like a signal. Everyone turns their head to look at you. He's from there!

That's how it was in the beginning. We didn't just lose a town, we lost our whole lives. We left on the third day. The reactor was on fire. I remember one of my friends saying, "It smells of reactor." It was an indescribable smell. But the papers were already writing about that. They turned Chernobyl into a house of horrors, although actually they just turned it into a cartoon. I'm only going to tell about what's really mine. My own truth.

It was like this: They announced over the radio that you couldn't take your cats. So we put her in the suitcase. But she didn't want to go, she climbed out. Scratched everyone. You can't take your belongings! All right, I won't take all my belongings, I'll take just one belonging. Just one! I need to take my door off the apartment and take it with me. I can't leave the door. I'll cover the entrance with some boards. Our door—it's our talisman, it's a family relic. My father lay on this door. I don't know whose tradition this is, it's not like that everywhere, but my mother told me that the deceased must be placed to lie on the door of his home. He lies there until they bring the coffin. I sat by my father all night, he lay on this door. The house was open. All night. And this door has little etch-marks on it. That's me growing up. It's marked there: first grade, second grade. Seventh. Before the army. And next to that: how my son grew. And my daughter. My whole life is written down on this door. How am I supposed to leave it?

I asked my neighbor, he had a car: "Help me." He gestured toward his head, like, You're not quite right, are you? But I took it with me, that door. At night. On a motorcycle. Through the woods. It was two years later, when our apartment had already been looted and emptied. The police were chasing me. "We'll shoot! We'll shoot!" They thought I was a thief. That's how I stole the door from my own home.

I took my daughter and my wife to the hospital. They had black spots all over their bodies. These spots would appear, then disappear. About the size of a five-kopek coin. But nothing hurt. They did some tests on them. I asked for the results. "It's not for you," they said. I said, "Then for who?"

Back then everyone was saying: "We're going to die, we're going to die. By the year 2000, there won't be any Belarussians left." My daughter was six years old. I'm putting her to bed, and she whispers in my ear: "Daddy, I want to live, I'm still little." And I had thought she didn't understand anything.

Can you picture seven little girls shaved bald in one room? There were seven of them in the hospital room . . . But enough! That's it! When I talk about it, I have this feeling, my heart tells me—you're betraying them. Because I need to describe it like I'm a stranger. My wife came home from the hospital. She couldn't take it. "It'd be better for her to die than to suffer like this. Or for me to die, so that I don't have to watch anymore." No, enough! That's it! I'm not in any condition. No.

We put her on the door . . . on the door that my father lay on. Until they brought a little coffin. It was small, like the box for a large doll.

I want to bear witness: my daughter died from Chernobyl. And they want us to forget about it.

*Nikolai Fomich Kalugin, father*


## MONOLOGUES BY THOSE WHO RETURNED

*The village of Bely Bereg, in the Narovlyansk region, in the Gomel oblast.*

*Speaking: Anna Pavlovna Artyushenko, Eva Adamovna Artyushenko, Vasily Nikolaevich Artyushenko, Sofya Nikolaevna Moroz, Nadezhda*

*Borisovna Nikolaenko, Aleksandr Fedorosvich Nikolaenko, Mikhail
Martynovich Lis.*

"And we lived through everything, survived everything . . ."

"Oh, I don't even want to remember it. It's scary. They chased
us out, the soldiers chased us. The big military machines rolled
in. The all-terrain ones. One old man—he was already on the
ground. Dying. Where was he going to go? 'I'll just get up,'
he was crying, 'and walk to the cemetery. I'll do it myself.'
What'd they pay us for our homes? What? Look at how pretty
it is here! Who's going to pay us for this beauty? It's a resort
zone!"

"Planes, helicopters—there was so much noise. The trucks with
trailers. Soldiers. Well, I thought, the war's begun. With the
Chinese or the Americans."

"My husband came home from the kolkhoz meeting, he says,
'Tomorrow we get evacuated.' And I say: 'What about the
potatoes? We didn't dig them out yet. We didn't get a chance.'
Our neighbor knocks on the door, and we sit down for a
drink. We have a drink and they start cursing the kolkhoz
chairman. 'We're not going, period. We lived through the
war, now it's radiation.' Even if we have to bury ourselves,
we're not going!"

"At first we thought, we're all going to die in two to three
months. That's what they told us. They propagandized us.
Scared us. Thank God—we're alive."

"Thank God! Thank God!"

"No one knows what's in the other world. It's better here. More familiar."

"We were leaving—I took some earth from my mother's grave, put it in a little sack. Got down on my knees: 'Forgive us for leaving you.' I went there at night and I wasn't scared. People were writing their names on the houses. On the wood. On the fences. On the asphalt."

"The soldiers killed the dogs. Just shot them. Bakh-bakh! After that I can't listen to something that's alive and screaming."

"I was a brigade leader at the kolkhoz. Forty-five years old. I felt sorry for people. We took our deer to Moscow for an exhibition, the kolkhoz sent us. We brought a pin back and a red certificate. People spoke to me with respect. 'Vasily Nikolaevich. Nikoleavich.' And who am I here? Just an old man in a little house. I'll die here, the women will bring me water, they'll heat the house. I felt sorry for people. I saw women walking from the fields at night singing, and I knew they wouldn't get anything. Just some sticks on payday. But they're singing . . ."

"Even if it's poisoned with radiation, it's still my home. There's no place else they need us. Even a bird loves its nest . . ."

"I'll say more: I lived at my son's on the seventh floor. I'd come up to the window, look down, and cross myself. I thought I heard a horse. A rooster. I felt terrible. Sometimes I'd dream about my yard: I'd tie the cow up and milk it and milk it. I wake up. I don't want to get up. I'm still there. Sometimes I'm here, sometimes there."

"During the day we lived in the new place, and at night we lived at home—in our dreams."

"The nights are very long here in the winter. We'll sit, sometimes, and count: who's died?"

"My husband was in bed for two months. He didn't say anything, didn't answer me. He was mad. I'd walk around the yard, come back: 'Old man, how are you?' He looks up at my voice, and that's already better. As long as he was in the house. When a person's dying, you can't cry. You'll interrupt his dying, he'll have to keep struggling. I took a candle from the closet and put it in his hand. He took it and he was breathing. I can see his eyes are dull. I didn't cry. I asked for just one thing: 'Say hello to our daughter and to my dear mother.' I prayed that we'd go together. Some gods would have done it, but He didn't let me die. I'm alive . . ."

"Girls! Don't cry. We were always on the front lines. We were Stakhanovites. We lived through Stalin, through the war! If I didn't laugh and comfort myself, I'd have hanged myself long ago."

"My mother taught me once—you take an icon and turn it upside-down, so that it hangs like that three days. No matter where you are, you'll always come home. I had two cows and two calves, five pigs, geese, chicken. A dog. I'll take my head in my hands and just walk around the yard. And apples, so many apples! Everything's gone, all of it, like that, gone!"

"I washed the house, bleached the stove. You need to leave some bread on the table and some salt, a little plate and three spoons.

As many spoons as there are souls in the house. All so we could come back."

"And the chickens had black cockscombs, not red ones, because of the radiation. And you couldn't make cheese. We lived a month without cheese and cottage cheese. The milk didn't go sour—it curdled into powder, white powder. Because of the radiation."

"I had that radiation in my garden. The whole garden went white, white as white can be, like it was covered with something. Chunks of something. I thought maybe someone brought it from the forest."

"We didn't want to leave. The men were all drunk, they were throwing themselves under cars. The big Party bosses were walking to all the houses and begging people to go. Orders: 'Don't take your belongings!' "

"The cattle hadn't had water in three days. No feed. That's it! A reporter came from the paper. The drunken milkmaids almost killed him."

"The chief is walking around my house with the soldiers. Trying to scare me: 'Come out or we'll burn it down! Boys! Give me the gas can.' I was running around—grabbing a blanket, grabbing a pillow."

"During the war you hear the guns all night hammering, rattling. We dug a hole in the forest. They'd bomb and bomb. Burned everything—not just the houses, but the gardens, the cherry trees, everything. Just as long as there's no war. That's what I'm scared of."

"They asked the Armenian broadcaster: 'Maybe there are Chernobyl apples?' 'Sure, but you have to bury the core really deep.'"

"They gave us a new house. Made of stone. But, you know, we didn't hammer in a single nail in seven years. It wasn't ours. It was foreign. My husband cried and cried. All week he works on the kolkhoz on the tractor, waits for Sunday, then on Sunday he lies against the wall and wails away . . ."

"No one's going to fool us anymore, we're not moving anywhere. There's no store, no hospital. No electricity. We sit next to a kerosene lamp and under the moonlight. And we like it! Because we're home."

"In town my daughter-in-law followed me around the apartment and wiped down the door handle, the chair. And it was all bought with my money, all the furniture and the Zhiguli, too, with the money the government gave me for the house and the cow. As soon as the money's finished, Mom's not needed anymore."

"Our kids took the money. Inflation took the rest. You can buy a kilo of nice candy with the money they gave us for our homes, although maybe now it wouldn't be enough."

"I walked for two weeks. I had my cow with me. They wouldn't let me in the house. I slept in the forest."

"They're afraid of us. They say we're infectious. Why did God punish us? He's mad? We don't live like people, we don't live according to His laws anymore. That's why people are killing one another."

"My nephews would come during the summer. The first summer they didn't come, they were afraid. But now they come. They take food, too, whatever you give them. 'Grandma,' they say, 'did you read the book about Robinson Crusoe?' He lived alone like us. Without people around. I brought half a pack of matches with me. An axe and a shovel. And now I have lard, and eggs, and milk—it's all mine. The only thing is sugar—can't plant that. But we have all the land we want! You can plow 100 hectares if you want. And no government, no bosses. No one gets in your way."

"The cats came back with us too. And the dogs. We all came back together. The soldiers didn't want to let us in. The riot troops. So at night—through the forest—like the partisans."

"We don't need anything from the government. Just leave us alone, is all we want. We don't need a store, we don't need a bus. We walk to get our bread. Twenty kilometers. Just leave us alone. We're all right by ourselves."

"We came back all together, three families. And everything here is looted: the stove is smashed, the windows, they took the doors off. The lamps, light switches, outlets—they took everything. Nothing left. I put everything back together with these hands. How else!"

"When the wild geese scream, that means spring is here. Time to sow the fields. And we're sitting in empty houses. At least the roofs are solid."

"The police were yelling. They'd come in cars, and we'd run into the forest. Like we did from the Germans. One time they came with the prosecutor, he huffed and puffed, they were going to put us up on Article 10. I said: 'Let them give me a year in jail. I'll serve it and come back here.' Their job is to yell, ours is to stay quiet. I have a medal—I was the best harvester on the kolkhoz. And he's scaring me with Article 10."

"Every day I'd dream of my house. I'm coming back to it: digging in the garden, or making my bed. And every time I find something: a shoe, or a little chick. And everything was for the best, it made me happy. I'd be home soon . . ."

"At night we pray to God, during the day to the police. If you ask me, 'Why are you crying?' I don't know why I'm crying. I'm happy to be living in my own house."

"We lived through everything, survived everything . . ."

"I got in to see a doctor. 'Sweety,' I say, 'my legs don't move. The joints hurt.' 'You need to give up your cow, grandma. The milk's poisoned.' 'Oh, no,' I say, 'my legs hurt, my knees hurt, but I won't give up the cow. She feeds me.'"

"I have seven children. They all live in cities. I'm alone here. I get lonely, I'll sit under their photographs. I'll talk a little. Just by myself. All by myself. I painted the house myself, it took six cans of paint. And that's how I live. I raised four sons and three daughters. And my husband died young. Now I'm alone."

"I met a wolf one time. He stood there, I stood there. We looked at each other. He went over to the side of the road, and I ran. My hat rose up I was so scared."

"Any animal is afraid of a human. If you don't touch him, he'll walk around you. Used to be, you'd be in the forest and you'd hear human voices, you'd run toward them. Now people hide from one another. God save me from meeting a person in the forest!"

"Everything that's written in the Bible comes to pass. It's written there about our kolkhoz, too. And about Gorbachev. That there'll be a big boss with a birthmark and that a great empire will crumble. And then the Day of Judgment will come. Everyone who lives in cities, they'll die, and one person from the village will remain. This person will be happy to find a human footprint! Not the person himself, but just his footprints."

"We have a lamp for light. A kerosene lamp. Ah-a. The women already told you. If we kill a wild boar, we take it to the basement or bury it ourselves. Meat can last for three days underground. The vodka we make ourselves."

"I have two bags of salt. We'll be all right without the government! Plenty of logs—there's a whole forest around us. The house is warm. The lamp is burning. It's nice! I have a goat, a kid, three pigs, fourteen chickens. Land—as much as I want; grass—as much as I want. There's water in the well. And freedom! We're happy. This isn't a kolkhoz anymore, it's a commune. We need to buy another horse. And then we won't need anyone at all. Just one horsey."

"This one reporter said, We didn't just return home, we went back a hundred years. We use a hammer for reaping, and a sickle for mowing. We flail wheat right on the asphalt."

"During the war they burned us, and we lived underground. In bunkers. They killed my brother and two nephews. All told, in my family we lost seventeen people. My mom was crying and crying. There was an old lady walking through the villages, scavenging. 'You're mourning?' she asked my mom. 'Don't mourn. A person who gives his life for others, that person is holy.' And I can do anything for my Motherland. Only killing I can't do. I'm a teacher, and I taught my kids to love others. That's how I taught them: 'Good will always triumph.' Kids are little, their souls are pure."

"Chernobyl is like the war of all wars. There's nowhere to hide. Not underground, not underwater, not in the air."

"We turned off the radio right away. We don't know any of the news, but life is peaceful. We don't get upset. People come, they tell us the stories—there's war everywhere. And like that socialism is finished and we live under capitalism. And the Tsar is coming back. Is that true?"

"Sometimes a wild boar will come into the garden, sometimes a fox. But people only rarely. Just police."

"You should come see my house, too."

"And mine. It's been a while since I had guests."

"I cross myself and pray: Dear God! Two times the police came and broke my stove. They took me away on a tractor. And me, I came back! They should let people in—they'd all come crawling back on their knees. They scattered our sorrow all over the globe. Only the dead come back now. The dead are allowed to. But the living can only come at night, through the forest."

"Everyone's rearing to get back for the harvest. That's it. Everyone wants to have his own back. The police have lists of people they'll let back, but kids under eighteen can't come. People will come and they're so glad just to stand next to their house. In their own yard next to the apple tree. At first they'll go cry at the cemetery, then they go to their yards. And they cry there, too, and pray. They leave candles. They hang them on their fences. Like on the little fences at the cemetery. Sometimes they'll even leave a wreath at the house. A white towel on the gate. The old woman reads a prayer: 'Brothers and sisters! Have patience!' "

"People take eggs, and rolls, and whatever else, to the cemetery. Everyone sits with their families. They call them: 'Sis, I've come to see you. Come have lunch.' Or: 'Mom, dear mom. Dad, dead dad.' They call the souls down from heaven. Those who had people die this year cry, and those whose people died earlier, don't. They talk, they remember. Everyone prays. And those who don't know how to pray, also pray."

"The only time I don't cry is at night. You can't cry about the dead at night. When the sun goes down, I stop crying. Remember their souls, oh Lord. And let their kingdom come."

"If you don't play, you lose. There was a Ukrainian woman at the market selling big red apples. 'Come get your apples! Chernobyl apples!' Someone told her not to advertise that, no one will buy them. 'Don't worry!' she says. 'They buy them anyway. Some need them for their mother-in-law, some for their boss.' "

"There was one guy, he came back here from jail. Under the amnesty. He lived in the next village. His mother died, the house was buried. He came over to us. 'Lady, give me some bread and some lard. I'll chop wood for you.' He gets by."

"The country is a mess—and people come back here. They run from the others. From the law. And they live alone. Even strangers. They're tough, there's no friendliness in their eyes. If they get drunk, they're liable to burn something down. At night we sleep with axes and pitchforks under our beds. In the kitchen next to the door, there's a hammer."

"There was a rabid fox here during the spring—when they're rabid they become tender, real tender. But they can't look at water. Just put a bucket of water in your yard, and you're fine. She'll run away."

"There's no television. No movies. There's one thing to do— look out the window. Well, and to pray, of course. There used to be Communism instead of God, but now there's just God. So we pray."

"We're people who've served our time. I'm a partisan, I was with the partisans a year. And when we beat back the Germans, I was on the front. I wrote my name on the Reichstag: Artyushenko. I took off my overcoat to build Communism. And where is this Communism?"

"We have Communism here—we live like brothers and sisters . . ."

"The year the war started, there weren't any mushrooms or any berries. Can you believe that? The earth itself felt the catastrophe. 1941. Oh, how I remember it! I've never forgotten the war. There was a rumor that they'd brought over all the POWs, if you recognized yours you could take him. All our women ran over! That night some brought home their men, and others brought home other men. But there was one scoundrel . . . He lived like everyone else, he was married, had two kids—he told the commandant that we'd taken in Ukrainians. Vasko, Sashko. The next day the Germans come on their motorcycles. We beg them, we get down on our knees. But they took them out of the village and shot them with their automatics. Nine men. And they were young, they were so good! Vasko, Sashko . . ."

"The boss-men come, they yell and yell, but we're deaf and mute. And we've lived through everything, survived everything . . ."

"But I'm talking about something else—I think about it a lot. At the cemetery. Some people pray loudly, others quietly. And some people say: 'Open up, yellow sand. Open up, dark night.' The forest might do it, but the sand never will. I'll ask gently: 'Ivan. Ivan, how should I live?' But he doesn't answer me anything, one way or the other."

"I don't have my own to cry about, so I cry about everyone. For strangers. I'll go to the graves, I'll talk to them."

"I'm not afraid of anyone—not the dead, not the animals, no one. My son comes in from the city, he gets mad at me. 'Why are you sitting here! What if some looter tries to kill you?' But what would he want from me? There's some pillows. In a simple house, pillows are your main furniture. If a thief tries to come in, the minute he peaks his head through the window, I'll chop it off with the axe. That's how we do it here. Maybe there is no God, or maybe there's someone else, but there's someone up there. And I'm alive."

"Why did that Chernobyl break down? Some people say, It was the scientists' fault. They grabbed God by the beard, and now he's laughing. But we're the ones who pay for it."

"We never did live well. Or in peace. We were always afraid. Just before the war they'd grab people. They came in black cars and took three of our men right off the fields, and they still haven't returned. We were always afraid."

"But now we're free. The harvest is rich. We live like barons."

"The only thing I have is a cow. I'd hand her in, if only they don't make another war. How I hate war!"

"Here we have the war of wars—Chernobyl."

"And the cuckoo is cuckooing, the magpies are chattering, roes are running. Will they reproduce—who knows? One morning I looked out in the garden, the boars were digging. They were wild. You can resettle people, but the elk and the boar, you can't. And water doesn't listen to borders, it goes along the earth, and under the earth."

"It hurts, girls. Oh, it hurts! Let's be quiet. They bring your coffin quietly. Careful. Don't want to bang against the door or the bed, don't want to touch anything or knock it over. Otherwise you have to wait for the next dead person. Remember their souls, oh Lord. May their kingdom come.

And let prayers be said for them where they're buried. We have everything here—graves. Graves everywhere. The dump trucks are working, and the bulldozers. The houses are falling. The gravediggers are toiling away. They buried the school, the headquarters, the baths. It's the same world, but the people are different. One thing I don't know is, Do people have souls? What kind? And how do they all fit in the next world? My grandpa took two days to die, I was hiding behind the stove and waiting: how's it going to fly out of his body? I went to milk the cow—I come back in, call him, he's lying there with his eyes open. His soul fled already. Or did nothing happen? And then how will we meet?"

"One old woman, she promises that we're immortal. We pray. Oh Lord, give us the strength to survive the weariness of our lives."

### MONOLOGUE ABOUT WHAT RADIATION LOOKS LIKE

My first scare was—some mornings in the garden and the yard we'd find these strangled moles. Who strangled them? Usually they don't come out from underground. Something was chasing them out. I swear on the Cross!

My son calls from Gomel: "Are the May bugs out?"

"No bugs, there aren't even any maggots. They're hiding."

"What about worms?"

"If you'd find a worm in the rain, your chicken'd be happy. But there aren't any."

"That's the first sign. If there aren't any May bugs and no worms, that means strong radiation."

"What's radiation?"

"Mom, that's a kind of death. Tell Grandma you need to leave. You'll stay with us."

"But we haven't even planted the garden."

If everyone was smart, then who'd be the dumb ones? It's on fire—so it's on fire. A fire is temporary, no one was scared of it then. They didn't know about the atom. I swear on the Cross! And we were living next door to the nuclear plant, thirty kilometers as the bird flies, forty on the highway. We were satisfied. You could buy a ticket and go there—they had everything, like in Moscow. Cheap salami, and always meat in the stores. Whatever you want. Those were good times!

Sometimes I turn on the radio. They scare us and scare us with the radiation. But our lives have gotten better since the radiation came. I swear! Look around: they brought oranges, three kinds of salami, whatever you want. And to the village! My grandchildren have been all over the world. The littlest just came back from France, that's where Napoleon attacked from once—"Grandma, I saw a pineapple!" My nephew, her brother, they took him to Berlin for the doctors. That's where Hitler started from on his tanks. It's a new world. Everything's different. Is that the radiation's fault, or what?

What's it like, radiation? Maybe they show it in the movies? Have you seen it? Is it white, or what? What color is it? Some people say it has no color and no smell, and other people say that it's black. Like earth. But if it's colorless, then it's like God. God is everywhere, but you can't see Him. They scare us! The apples are hanging in the garden, the leaves are on the trees, the potatoes are in the fields. I don't think there was any Chernobyl, they made it up. They tricked people. My sister left with her husband. Not far from here, twenty kilometers. They lived there two months, and the neighbor comes running: "Your cow sent radiation to my cow! She's falling down."

"How'd she send it?" "Through the air, that's how, like dust. It flies." Just fairy tales! Stories and more stories.

But here's what did happen. My grandfather kept bees, five nests of them. They didn't come out for two days, not a single one. They just stayed in their nests. They were waiting. My grandfather didn't know about the explosion, he was running all over the yard: what is this? What's going on? Something's happened to nature. And their system, as our neighbor told us, he's a teacher, it's better than ours, better tuned, because they heard it right away. The radio wasn't saying anything, and the papers weren't either, but the bees knew. They came out on the third day. Now, wasps—we had wasps, we had a wasps' nest above our porch, no one touched it, and then that morning they weren't there anymore—not dead, not alive. They came back six years later. Radiation: it scares people and it scares animals. And birds. And the trees are scared, too, but they're quiet. They won't say anything. It's one big catastrophe, for everyone. But the Colorado beetles are out and about, just as they always were, eating our potatoes, they scarf it down to the leaf, they're used to poison. Just like us.

But if I think about it—in every house, someone's died. On that street, on the other side of the river—all the women are without men, there aren't any men, all the men are dead. On my street, my grandfather's still alive, and there's one more. God takes the men earlier. Why? No one can tell us. But if you think about it—if only the men were left, without any of us, that wouldn't be any good either. They drink, oh do they drink! From sadness. And all our women are empty, their female parts are ruined in one in three of them, they say. In the old and the young, too. Not all of them managed to give birth in time. If I think about it—it just went by, like it never was.

What else will I say? You have to live. That's all.

And also this. Before, we churned our butter ourselves, our cream, made cottage cheese, regular cheese. We boiled milk dough. Do they eat that in town? You pour water on some flour and mix it in, you get these torn bits of dough, then you put these in the pot with some boiled water. You boil that and pour in some milk. My mom showed it to me and she'd say: "And you, children, will learn this. I learned it from my mother." We drank juice from birch and maple trees. We steamed beans on the stove. We made sugared cranberries. And during the war we gathered stinging-nettle and goose-foot. We got fat from hunger, but we didn't die. There were berries in the forest, and mushrooms. But now that's all gone. I always thought that what was boiling in your pot would never change, but it's not like that. You can't have the milk, and the beans either. They don't let you eat the mushrooms or the berries. They say you have to put the meat in water for three hours. And that you have to pour off the water twice from the potatoes when you're boiling them. Well, you can't wrestle with God. You have to live. They scare us, that even our water you can't drink. But how can you do without water? Every person has water inside her. There's no one without water. Even rocks have water in them. So, maybe, water is eternal? All life comes from water. Who can you ask? No one will say. People pray to God, but they don't ask him. You just have to live.

*Anna Petrovna Badaeva, re-settler*


## MONOLOGUE ABOUT A SONG WITHOUT WORDS

I'll get down on my knees to beg you—please, find our Anna Sushko. She lived in our village. In Kozhushki. Her name is Anna Sushko. I'll tell you how she looked, and you'll type it up. She has a hump, and she was mute from birth. She lived by herself. She was sixty. During the time of the transfer they put her in an ambulance and drove her off somewhere. She never learned how to read, so we never got any letters from her. The lonely and the sick were put in special places. They hid them. But no one knows where. Write this down . . .

The whole village felt sorry for her. We took care of her, like she was a little girl. Someone would chop wood for her, someone else would bring milk. Someone would sit in the house with her of an evening, heat the stove. Two years we all lived in other places, then we came back to our houses. Tell her that her house is still there. The roof is still there, the windows. Everything that's broken or been stolen, we can fix. If you just tell us her address, where she's living and suffering, we'll go there and bring her back. So that she won't die of sorrow. I beg you. An innocent spirit is suffering among strangers . . .

There's one other thing about her, I forgot. When something hurts, she sings this song. There aren't any words, it's just her voice. She can't talk. When something hurts, she just sings: a-a-a. It makes you feel sorry.

*Mariya Volchok, neighbor*


THREE MONOLOGUES ABOUT A HOMELAND

*Speaking: The K. family—mother and daughter, plus a man who doesn't speak a word (the daughter's husband).*

*Daughter:*
At first I cried day and night. I wanted to cry and talk. We're from Tajikistan, from Dushanbe. There's a war there.

I shouldn't be talking about this now. I'm expecting—I'm pregnant. But I'll tell you. They come onto the bus one day to check our passports. Just regular people, except with automatic weapons. They look through the documents and then push the men out of the bus. And then, right there, right outside the door, they shoot them. They don't even take them aside. I would never have believed it. But I saw it. I saw how they took out two men, one was so young, handsome, and he was yelling something at them. In Tajik, in Russian. He was yelling that his wife just gave birth, he has three little kids at home. But they just laughed, they were young, too, very young. Just regular people, except with automatic weapons. He fell. He kissed their sneakers. Everyone was quiet, the whole bus. Then we drove off, and we heard: ta-ta-ta. I was afraid to look back. [*Starts crying.*]

I'm not supposed to be talking about this. I'm expecting a baby. But I'll tell you. Just one thing, though: don't write my last name. I'm Svetlana. We still have relatives there. They'll kill them. I used to think we'd never have any more wars. Such a big country, I thought, my beloved country. The biggest! During Soviet times they'd tell us that we were living poorly and humbly because there had been a big war, and the people suffered, but now that we have a mighty army, no one will ever touch us again. No one will defeat us! But then we started shooting one another. It's not a war like there used to be, like my grandfather remembered, he marched all the way to Germany. Now it's a neighbor shooting his neighbor, boys who went to school together, and now they kill each other, and rape girls that they sat next to in school. Everyone's gone crazy.

Our husbands are silent. The men here are silent. They won't say anything to you. People yelled at them as they were leaving, that they were running away just like women. That they were cowards, betraying their motherland. But is that bad?

Is it a bad thing not to be able to shoot? My husband is a Tajik, he was supposed to go and kill people. But he said: "Let's leave. I don't want to go to war. I don't need an automatic." That's his land, but he left, because he doesn't want to kill another Tajik, the same kind of Tajik as he is. But he's lonely here, his brothers are all still there, fighting. One already got killed. His mother lives there. His sisters. We rode here on the Dushanbe train, the windows were broken, it was cold and unheated. No one was shooting, but they threw rocks at the train, broke the windows. "Russians, get out! Occupiers! Quit robbing us!" But he's a Tajik, and he had to listen to all this. And our kids heard it. Our daughter was in first grade, she was in love with a boy, a Tajik. She came home from school: "Mom, what am I, a Tajik or Russian?" How do you explain?

I'm not supposed to be talking about this . . . but I'll tell you. The Pamir Tajiks are fighting the Kulyab Tajiks. They're all Tajiks, they have the same Koran, the same faith, but the Kulyabs kill the Pamirs, and the Pamirs kill the Kulyabs. First they'd go out into the city square, yelling, praying. I wanted to understand what was happening, so I went too. I asked one of the old men: "What are you protesting against?" They said: "Against the Parliament. They told us this was a very bad person, this Parliament." Then the square emptied and they started shooting. All of a sudden it became a different country, an unrecognizable country. The East! And before that we thought we were living on our own land. By Soviet laws. There are so many Russian graves there, but there's no one to cry at them. They graze livestock on the Russian cemeteries. And goats. Old Russian men wander around, going through trash cans . . .

I worked in a maternity ward as a nurse. I had night duty. This woman is giving birth, it's a difficult birth, and she's yelling—suddenly an orderly runs in, she's not wearing gloves, no robe. What's going on? To come into the maternity ward like that? "Girls, there are people! They're wearing masks, they have guns." Then they come in: "Give us the drugs! And the alcohol!" "There aren't any drugs or alcohol." They put the doctor up against the wall—give it here! And then the woman who's giving birth yells with relief, happily. And the baby starts crying, it's just-just come out. I lean over it to look, I can't even remember now whether it was a boy or a girl. It didn't have a name or anything yet. And these robbers say to us: what is it, a Kulyab or a Pamir? Not, boy or girl, but *Kulyab* or *Pamir*? We don't say anything. They start yelling: "What is it?" We don't say anything. So they grab the little baby, it's been on this earth for maybe five, ten minutes, and they throw it out the window. I'm a nurse, I'd never seen a baby die before. And here—I'm not supposed to remember this now. [*Starts crying.*] How are you supposed to live after that? How are you supposed to give birth? [*Cries.*]

After that, in the maternity ward, the skin started coming off my hands. My veins swelled up. And I was so indifferent to everything. I didn't want to get out of bed. [*Cries.*] I'd get to the hospital and then turn around. By then I was pregnant myself. I couldn't give birth there. So we came here. To Belarus. To Narovlya. Small, quiet town. And don't ask me anything else. I've told you everything. [*Cries.*] Wait. I want you to know. I'm not afraid of God. I'm afraid of man. At first we asked people: "Where is the radiation?" "See where you're standing? That's where it is." So it's everywhere? [*Cries.*] There are many empty houses. People left. They were scared.

But I'm not scared here the way I was there. We were left without a homeland, we're no-one's. The Germans all went back to Germany, the Tatars to the Crimea, when they were allowed to, but no one needs Russians. What are we supposed to hope for? What do we wait for? Russia never saved its people, because it's so big, it's endless. And to be honest, I don't feel like Russia is my homeland. We were raised differently, our homeland is the Soviet Union. Now it's impossible to know how you're supposed to save yourself. At least here no one's playing with guns, and that's good. Here they gave us a house, and they gave my husband a job. We wrote a letter to our friends back home, and they came yesterday. For good. They came at night and they were afraid to come out of the train station, they stayed there all night, sitting on their suitcases, not letting their kids out. And then they see: people are walking down the street, laughing, smoking. They showed them our street, escorted them right to our house. They couldn't believe it, because back there we stopped living normal lives. Here they got up in the morning and went to the store, they saw butter, and cream—and right there, in the store, they told us this themselves, they bought five bottles of cream and drank them right there. People were looking at them like they were crazy. But they hadn't seen cream or butter in two years. You can't buy bread in Tajikistan. There's a war. It's impossible to explain to someone who hasn't seen what it's like.

My soul was dead there. I would have given birth to something without a soul. There aren't many people here, and the houses are empty. We live near the forest. I don't like it when there are a lot of people. Like at the train station. Or during the war. [*Breaks into tears completely and stops talking.*]

*Mother:*
The war—that's the only thing I can talk about. Why did we come here? To Chernobyl? Because no one's going to chase us out of here. No one will kick us off this land. It's not anyone's land now. God took it back. People left it.

In Dushanbe I was deputy chief of the train station. There was one other deputy, a Tajik. Our kids grew up together, went to school, we all got together on the holidays: New Year's, May Day. We drank beer together, ate *plof* together. He'd call me "sister, my sister, my Russian sister." And then one day he comes in, we sat in the same office, and he stops in front of my desk and yells:

"When are you going back to your Russia, huh? This is our land!"

I thought I'd go crazy. I jumped up at him.

"Where's your coat from?"

"Leningrad," he said. He was surprised.

"Take off that Russian coat, you son-of-a-bitch!" And I tore the coat off him. "Where's your hat from? You bragged to me they sent it from Siberia! Off with it, you! And the shirt! The pants! Those were made in Moscow! They're Russian, too!"

I'd have stripped him to his underwear. He was a big guy, I came up to his shoulder, but I'd have torn everything off him. People were already gathering around. He's crying: "Get away from me, you're crazy!"

"No, give me back everything that's mine, that's Russian! I'll take it all!"

I almost went crazy.

"Give me your socks! Your shoes!"

We worked at night and during the day. Trains were leaving overfilled. People were running. Many Russians left—thousands, tens of thousands. There's still one Russia. I see the Moscow train off at two in the morning, and there are still some kids in the hall from the town of Kurgan-Tyube, they didn't make it to the train. I covered them up, I hid them. Two men come over to me, they've got automatics.

"Oh, boys, what are you doing here?" Meanwhile my heart's beating.

"It's your own fault, all your doors are wide open."

"I was sending off a train. I didn't get a chance to close them."

"Who are those kids over there?"

"Those are ours, from Dushanbe."

"Maybe they're from Kurgan? They're Kulyabs?"

"No, no. They're ours."

So they left. And if they'd opened the hall? They'd have . . . And me, too, while they were at it, a bullet to the head. There's only one government there—the man with the gun. In the morning I put the kids on the train to Astrakhan, I told the conductors to transport them like they do watermelons, to not open the door. [*Silent. Then cries for a long time.*] Is there anything more frightening than people? [*Silent again.*]

One time, when I was here already, I was walking down the street and I started looking back, because I thought someone was following me. Not a day went by there when I didn't think of death. I always left the house wearing clean clothes, a freshly laundered blouse, skirt, underthings. Just in case I got killed. Now I walk through the forest by myself and I'm not afraid of anyone. There aren't any people in the forest, not a soul. I walk and wonder whether all of that really happened to me or not? Sometimes I'll run into some hunters: they have rifles, a dog, and a dosimeter. They also have guns, but they're not like the others, they don't hunt people. If I hear gunfire, I know they're shooting some crows or chasing off a rabbit. [*Silent.*] So I'm not scared here. I can't be afraid of the earth, the water. I'm afraid of people. Over there he goes to the market and buys an automatic weapon for a hundred dollars.

I remember one guy, a Tajik, I saw him chasing this other guy. He was chasing another person! The way he was running, the way he was breathing, I could tell he wanted to kill him. But the other one got away. He hid. And this one comes back, he walks past me and says, "Ma'am, where do I get some water around here?" He's so casual about it, like nothing happened. We had a bucket of water at the station, I showed it to him. Then I looked him in the eye and I said: "Why are you chasing one another? Why are you killing?" And he looked like he felt ashamed. "All right, ma'am, not so loud." But when they're together, they're different. If there'd been three of them, or even two, they'd have put me up against the wall. When you're one-on-one you can still talk to a person.

We got to Tashkent from Dushanbe, but we had to go further, to Minsk. There weren't any tickets—none! It's very clever the way they have it set up, until you've given someone a bribe and you're on the plane, there are endless problems: it's too heavy, or too much volume, you can't have this, you have to put that away. They made me put everything on the scale twice, until I realized what was happening and gave them some money. "Should have done that from the start, instead of arguing so much." Everything's so simple! Our container, it weighed two tons, they made us unload it. "You're coming from a war zone, maybe you've got some firearms in there? Marijuana?" They kept us there two nights. I went to the station boss but in the waiting room I met a good woman, she explained things to me: "You won't get anywhere here, you'll demand fairness, meanwhile they'll throw your container in a field and take everything you own." So what do we do? We spent the whole night picking through it: clothes, some mattresses, an old refrigerator, two bags of books. "You're shipping valuable books?" We looked: Chernyshevsky's *What Is to Be Done?*, Sholokhov's *Virgin Soil Upturned*. We laughed. "How many refrigerators do you have?" "Just one, and that one's been broken." "Why didn't you bring declarations?" "How were we supposed to know? It's the first time we've run away from a war." We lost two homelands at once—Tajikistan and the Soviet Union.

I walk through the forest and think. Everyone else is always watching television—what's happening there? How is everyone? But I don't want to.

We had a life . . . a different life. I was considered an important person, I had a military rank, lieutenant colonel of train-based troops. Here I was unemployed until I found work cleaning up at the town council. I wash the floors. This life has passed, and I don't have enough strength for another. Some people here feel sorry for us, others are unhappy—"the refugees are stealing the potatoes, they dig them up at night." My mother said that during the big war people felt sorry for each other more. Recently they found a horse in the forest that had gone wild. It was dead. In another place they found a rabbit. They hadn't been killed, but they were dead. This made everyone worried. But when they found a dead bum, no one worried about that. For some reason everyone's grown used to dead people.

*Lena M.—from Kyrgyzstan. She sits at the entrance to her home as if posing for a photograph. Her five children sit near her, as does their cat, Metelitsa, whom they brought with them:*
We left like we were leaving a war. We grabbed everything, and the cat followed us to the train station, so we took him, too. We were on the train for twelve days. The last two days all we had left was some canned cabbage salad and boiled water. We guarded the door—with a crowbar, and an axe, and a hammer. I'll put it this way—one night some looters attacked us. They almost killed us. They'll kill you now

for a television or refrigerator. It was like we were leaving a war, although they're not shooting yet in Kyrgyzstan. There were massacres, even under Gorbachev, in Osh, the Kyrgyz and the Uzbeks—but it settled down somehow. But we're Russian, though the Kyrgyz are afraid of it too. You'd be in line for bread and they'd start yelling, "Russians, go home! Kyrgyzstan for the Kyrgyz!" And they'd push you out of line. And then they'd add something in Kyrgyz, like, Here we are, there's not even enough bread for us, and we have to feed them? I don't really know their language very well, I just learned a few words so I could haggle at the market, buy something.

We had a motherland, and now it's gone. What am I? My mother's Ukrainian, my father's Russian. I was born and raised in Kyrgyzstan, and I married a Tatar. So what are my kids? What is their nationality? We're all mixed up, our blood is all mixed together. On our passports, my kids and mine, it says "Russian," but we're not Russian. We're Soviet! But that country—where I was born—no longer exists. The place we called our motherland doesn't exist, and neither does that time, which was also our motherland. We're like bats now. I have five children. The oldest is in eighth grade, and the youngest is in kindergarten. I brought them here. Our country no longer exists, but we do.

I was born there. I grew up there. I helped build a factory, then I worked at the factory. "Go back where you're from; this is all ours." They didn't let me take anything but my kids. "This is all ours." And where is mine? People are fleeing. All the Russians are. The Soviets. No one needs them, and no one is waiting for them.

And I was happy once. All my children were born of love. I gave birth like this: boy, boy boy, and then girl,

girl. I don't want to talk anymore. I'll start crying. [*But she adds a bit more.*] We'll wait in Chernobyl. This is our home now. Chernobyl is our home, our motherland. [*She smiles suddenly.*] The birds here are the same as everywhere. And there's still a Lenin statue. [*When we're already at the gate, saying goodbye, she says some more.*] Early one morning the neighbors are hammering away on the house, taking the boards off the windows. I see a woman. I say, "Where are you from?" "From Chechnya." She doesn't say anything more, just starts crying . . .

People ask me, they're surprised, they don't understand. "Why are you killing your children?" Oh, God, where do you find the strength to meet the things that the next day is going to bring? I'm not killing them, I'm saving them. Here I am, forty years old and completely gray. And they're surprised. They don't understand. They say: "Would you bring your kids to a place where there was cholera or the plague?" But that's the plague and that's cholera. This fear that they have here in Chernobyl, I don't know about it. It's not part of my memory.

## MONOLOGUE ABOUT HOW A PERSON IS ONLY CLEVER AND REFINED IN EVIL

I was running away from the world. At first I hung around train stations, I liked it there, so many people and you're all by yourself. Then I came here. Freedom is here.

I've forgotten my own life. Don't ask me about it. I remember what I've read in books, and what other people have told me, but my own life I've forgotten. It was a long time ago. I did wrong. But there's no sin that God won't forgive if the penance is sincere.

A man can't possibly be happy. He's not supposed to be. God saw that Adam was lonely and gave him Eve. For happiness, not for sin. But man isn't capable of happiness. Like me, for example, I don't like twilight. I don't like the dark. This corridor, like right now, between light and dark. I still don't understand where I was—how it was—and it doesn't matter. I can live or not live, it doesn't matter. The life of man is like grass: it blossoms, dries out, and then goes into the fire. I fell in love with contemplation. Here you can die equally well from an animal or from the cold. There's no one for tens of kilometers. You can chase off demons by fasting and praying. You fast for your flesh, and you pray for your soul. But I'm never lonely, a man who believes can never be lonely. I ride around the villages—I used to find spaghetti, flour—even vegetable oil. Canned fruit. Now I go to the cemeteries—people leave food and drink for the dead. But the dead don't need it. They don't mind. In the fields there's wild grain, and in the forest there are mushrooms and berries. Freedom is here.

I read in a book—it was by Father Sergei Bulgakov—"It's certain that God created the world, and therefore the world can't possibly fail," and so it's necessary to "endure history courageously and to the very end." Another thinker says, I don't remember his name, but in effect he says: "Evil is not an actual substance. It is the absence of good, in the way that darkness is simply the absence of light." It's easy to find books here. Now, an empty clay pitcher, or a spoon or fork, that you won't find, but books are all over. The other day I found a volume of Pushkin. "And the thought of death is sweet to my soul." I remembered that. Yes: "The thought of death." I am here alone. I think about death. I've come to like thinking. And silence helps you to prepare yourself. Man lives with death, but he doesn't understand what it is. But I'm here alone. Yesterday I chased a wolf and a she-wolf out of the school, they were living there.

Question: Is the world as it's depicted in words the real world? Words stand between the person and his soul.

And I'll say this: birds, and trees, and ants, they're closer to me now than they were. I think about them, too. Man is frightening. And strange. But I don't want to kill anyone here. I fish here, I have a rod. Yes. But I don't shoot animals. And I don't set traps. You don't feel like killing anyone here.

Prince Myshkin said: "Is it possible to see a tree and not be happy?" Yes . . . I like to think. Whereas man complains above all, instead of thinking.

What's the point of looking at evil? Evil is important, of course. Sin isn't a matter of physics. You have to acknowledge the nonexistent. It says in the Bible: "For those who walk in light, it is one way; and for the rest, there is the teaching." If you take a bird—or any other living thing—we can't understand them, because they live for themselves, and not for others. Yes. Everything around is fluid, to put it in just one word.

Everything that walks on four legs looks at the ground, heads for the ground. Only man stands up, and raises his hands and face to the sky. To prayer. To God. The old woman in the church prays, "To each of us according to our sins." But neither the scientist, nor the engineer, nor the soldier will admit to it. They think: "I have nothing to repent of. Why should I repent?" Yes . . .

I pray simply. I pray for myself. Oh Lord, call upon me! Hear me! Only in evil is a man clever and refined. But how simple and sympathetic he is when speaking the honest words of love. Even when the philosophers use words they are only approximations of the thoughts they have felt. The word corresponds exactly to what is in the soul only in prayer, in the thought of prayer. I feel that physically to be true. Oh, Lord, call upon me! Hear me!

And man, also.

I am afraid of man. And also I want to meet him. I want to meet a good person. Yes. Here it's either looters, who are hiding out, or people like me, martyrs.

What's my name? I don't have a passport. The police took it. They beat me. "What're you hanging around for?" "I'm not hanging around—I'm repenting." They beat me harder after that. They beat me on the head. So you should write: "God's servant Nikolai. Now a free man."

## SOLDIERS' CHORUS

*Artyom Bakhtiyarov, private; Oleg Leontyevich Vorobey, liquidator; Vasily Iosifovich Gusinovich, driver and scout; Gennady Viktorovich Demenev, police officer; Vitaly Borisovich Karbalevich, liquidator; Valentin Kmkov, driver and private; Eduard Borisovich Korotkov, helicopter pilot; Igor Litvin, liquidator; Ivan Aleksandrovich Lukashuk, private; Aleksandr Ivanovich Mikhalevich, Geiger operator; Major Oleg Leonodovich Pavlov, helicopter pilot; Anatoly Borisovich Rybak, commander of a guard regiment; Viktor Sanko, private; Grigory Nikolaevich Khvorost, liquidator; Aleksandr Vasilievich Shinkevich, police officer; Vladimir Petrovich Shved, captain; Aleksandr Mikhailovich Yasinskiy, police officer.*

Our regiment was given the alarm. It was only when we got to the Belorusskaya train station in Moscow that they told us where we were going. One guy, I think he was from Leningrad, began to protest. They told him they'd drag him before a military tribunal. The commander said exactly that before the troops: "You'll go to jail or be shot." I had other feelings, the complete opposite of that guy. I wanted to do something heroic. Maybe it was kid's stuff. But there were others like me. We had guys from all over the Soviet Union. Russians, Ukrainians, Kazakhs, Armenians . . . It was scary but also fun, for some reason.

So they brought us in, and they took us right to the power station. They gave us white robes and white caps. And gauze surgical masks. We cleaned the territory. We spent a day cleaning down below, and then a day above, on the roof of the reactor. Everywhere we used shovels. The guys who went up, we called them the storks. The robots couldn't do it, their systems got all crazy. But we worked. And we were proud of it.

*

We rode in—there was a sign that said, Zone Off Limits. I'd never been to war, but I got a familiar feeling. I remembered it from somewhere. From where? I connected it to death, for some reason . . .

We met these crazed dogs and cats on the road. They acted strange: they didn't recognize us as people, they ran away. I couldn't understand what was wrong with them until they told us to start shooting at them . . . The houses were all sealed up, the farm machinery was abandoned. It was interesting to see. There was no one, just us and the police on their patrols. You'd walk into a house—there were photographs on the wall, but no people. There'd be documents lying around: people's komsomol IDs, other forms of identification, awards. At one place we took a television for a while—we borrowed it, say—but as far as anyone actually taking something home with them, I didn't see that. First of all, because you sensed that these people would be back any minute. And second, this was connected somehow with death.

People drove to the block, the actual reactor. They wanted to photograph themselves there, to show the people at home.

They were scared, but also really curious: what was this thing? I didn't go, myself, I have a young wife, I didn't want to risk it, but the boys downed a few shots and went over. So . . . [*Silent.*]

*

The village street, the field, the highway—all of it without any people. A highway to nowhere. Electrical wires on the posts to nowhere. At first there were still lights on in the houses, but then they turned those off. We'd be driving around, and a wild boar would jump out of a school building at us. Or else a rabbit. Everywhere, animals instead of people: in the houses, the schools, the clubs. There are still posters: "Our goal is the happiness of all mankind." "The world proletariat will triumph." "The ideas of Lenin are immortal." You go back to the past. The kolkhoz offices have red flags, brand-new wimples, neat piles of printed banners with profiles of the great leaders. On the walls—pictures of the leaders; on the desks—busts of the leaders. A war memorial. A village churchyard. Houses that were shut up in a hurry, gray cement cowpens, tractor mechanic's shops. Cemeteries and victims. As if a warring tribe had left some base in a hurry and then gone into hiding.

We'd ask each other: is this what our life is like? It was the first time we saw it from the outside. The very first time. It made a real impression. Like a smack to the head. . . . There's a good joke: the nuclear half-life of a Kiev cake is thirty-six hours. So . . . And for me? It took me three years. Three years later I turned in my Party card. My little Red book. I became free in the Zone. Chernobyl blew my mind. It set me free.

*

There's this abandoned house. It's closed. There's a cat on the windowsill. I think—must be a clay cat. I come over, and it's a real cat. He ate all the flowers in the house. Geraniums. How'd he get in? Or did they leave him there?

There's a note on the door: "Dear kind person, Please don't look for valuables here. We never had any. Use whatever you want, but don't trash the place. We'll be back." I saw signs on other houses in different colors—"Dear house, forgive us!" People said goodbye to their homes like they were people. Or they'd written: "we're leaving in the morning," or, "we're leaving at night," and they'd put the date and even the time. There were notes written on school notebook paper: "Don't beat the cat. Otherwise the rats will eat everything." And then in a child's handwriting: "Don't kill our Zhulka. She's a good cat." [*Closes his eyes.*] I've forgotten everything. I only remember that I went there, and after that I don't remember anything. I forgot all of it. I can't count money. My memory's not right. The doctors can't understand it. I go from hospital to hospital. But this sticks in my head: you're walking up to the house, thinking the house is empty, and you open the door and there's this cat. That, and those kids' notes.

*

I was called on. My assignment was not to let any of the old inhabitants back into the evacuated villages. We set up roadblocks, built observation posts. They called us "partisans," for some reason. It's peacetime, and we're standing there, in military fatigues. The farmers didn't understand why, for example, they couldn't take a bucket from their yard, or a pitcher, saw, axe. Why they couldn't harvest the crops. How do you tell them? And in fact it was like this: on one side of the road there were soldiers, keeping people out, and on the other side cows were grazing, the harvesters were buzzing, the grain was being shipped. The old women would come and cry: "Boys, let us in. It's our land. Our houses." They'd bring eggs, bacon, homemade vodka. They cried over their poisoned land. Their furniture. Their things.

Your mind would turn over. The order of things was shaken. A woman would milk her cow, and next to her there'd be a soldier who had to make sure that when she was done milking, she'd pour the milk out on the ground. An old woman carries a basket of eggs, and next to her there's a soldier walking to make sure she buries them. The farmers were raising their precious potatoes, harvesting them really quietly, but in fact they had to be buried. The worst part was, the least comprehensible part, was that everything was so—beautiful! That was the worst. All around, it was just beautiful. I would never see such people again. Everyone's faces just looked crazy. Their faces did, and so did ours.

*

I'm a soldier. If I'm ordered to do something, I need to do it. But I felt this desire to be a hero, too. You were supposed to. The political workers gave speeches. There were items on the radio and television. Different people reacted differently: some wanted to be interviewed, show up on television, and some just saw it as their job, and then a third type—I met people like this, they felt they were doing heroic work. We were well paid, but it was as if that didn't matter. My salary was 400 rubles, whereas there I got 1000 (that's in those Soviet rubles). Later people said, "They got piles of money and now they come back and get the first cars, the first furniture sets." Of course it hurts. Because there was that heroic aspect, also.

I was scared before I went there. For a little while. But then when I got there the fear went away. It was all orders, work, tasks. I wanted to see the reactor from above, from a helicopter—I wanted to see what had really happened in there. But that was forbidden. On my medical card they wrote that I got 21 roentgen, but I'm not sure that's right. The procedure was very simple: you flew to the provincial capital, Chernobyl (which is a small provincial town, by the way, not something enormous, as I'd imagined), there's a man there with a dosimeter, 10-15 kilometers away from the power station, he measures the background radiation. These measurements would then be multiplied by the number of hours that we flew each day. But I would go from there to the reactor, and some days there'd be 80 roentgen, some days 120. Sometimes at night I'd circle over the reactor for two hours. We photographed it with infrared lighting, but the pieces of scattered graphite on the film were, like, radiated—you couldn't see them during the day.

I talked to some scientists. One told me, "I could lick your helicopter with my tongue and nothing would happen to me." Another said, "You're flying without protection? You don't want to live too long? Big mistake! Cover yourselves!" We lined the helicopter seats with lead, made ourselves some lead vests, but it turns out those protect you from one sets of rays, but not from another set. We flew from morning to night. There was nothing spectacular in it. Just work, hard work. At night we watched television—the World Cup was on, so we talked a lot about soccer.

We started thinking about it—I guess it must have been—three years later. One of the guys got sick, then another. Someone died. Another went insane and killed himself. That's when we started thinking. But we'll only really understand in about 20-30 years. For me, Afghanistan (I was there two years) and then Chernobyl (I was there three months), are the most memorable moments of my life.

I didn't tell my parents I'd been sent to Chernobyl. My brother happened to be reading *Izvestia* one day and saw my picture. He brought it to our mom. "Look," he says, "he's a hero!" My mother started crying.

*

We were driving, and you know what I saw? By the side of the road? Under a ray of light—this thin little sliver of light—something crystal. These . . . We were going in the direction of Kalinkovich, through Mozyr. Something glistened. We talked about it—in the village where we worked, we all noticed there were tiny little holes in the leaves, especially on the cherry trees. We'd pick cucumbers and tomatoes—and the leaves would have these black holes. We'd curse and eat them.

I went. I didn't have to go. I volunteered. At first you didn't see any indifferent people there, it was only later that you saw the emptiness in their eyes, when they got used to it. I was after a medal? I wanted benefits? Bullshit! I didn't need anything for myself. An apartment, a car—what else? Right, a dacha. I had all those things. But it exerted a sort of masculine charm. Manly men were going off to do this important thing. And everyone else? They can hide under women's skirts, if they want. There were guys with pregnant wives, others had little babies, a third had burns. They all cursed to themselves and came anyway.

We came home. I took off all the clothes that I'd worn there and threw them down the trash chute. I gave my cap to my little son. He really wanted it. And he wore it all the time.

Two years later they gave him a diagnosis: a tumor in his brain . . . You can write the rest of this yourself. I don't want to talk anymore.

\*

I had just come home from Afghanistan. I wanted to live a little, to get married. I wanted to get married right away. And suddenly here's this announcement with a red banner, "Special Call-Up," come to this address within the hour. Right away my mother started crying. She thought I was being called up again for the war.

Where are we going? Why? There was no information at all. At the Slutsk station, we changed, they gave us equipment, and then we were told that we were going to the Khoyniki regional center. We got to Khoyniki, and people there didn't know anything. They took us further, to a village, and there's a wedding going on: young people dancing, music, drinking vodka. Just a normal marriage. And we have an order: get rid of the topsoil to the depth of one spade.

On May 9, V-Day, a general came. They lined us up, congratulated us on the holiday. One of the guys got up the courage and asked, "Why aren't they telling us the radiation levels? What kind of doses are we getting?" Just one guy. Well, after the general left, the brigadier called him in and gave him hell. "That's a provocation! You're an alarmist!" A few days later they gave us some gas masks, but no one used them. They showed us dosimeters a couple of times, but they never actually handed them to us. Once every three months they let us go home for a few days. We had one goal then: to buy vodka. I lugged back two backpacks filled with bottles. The guys raised me up on their arms.

Before we went home we were called in to talk to a KGB man. He was very convincing in telling us we shouldn't talk to anyone, anywhere, about what we'd seen. When I made it back from Afghanistan, I knew that I'd live. Here it was the opposite: it'd kill you only after you got home.

*

What do I remember? What stuck in my memory?

I've spent all day riding through all the villages, measuring the radiation. And not one of the women offers me an apple. The men are less afraid: they'll come up to me and offer some vodka, some lard. Let's eat. It's awkward to turn them down, but then eating pure cesium doesn't sound so great, either. So I drink, but I don't eat.

But in one village they do sit me down at the table—grilled lamb and everything. The host gets a little drunk and admits it was a young lamb. "I had to slaughter him. I couldn't stand to look at him anymore. He was the ugliest damn thing! Almost makes me not want to eat him." Me: I just drink a whole glass of vodka real quick. After hearing that . . .

*

It was ten years ago. It's as if it never happened, and if I hadn't gotten sick, I'd have forgotten by now.

You have to serve the motherland! Serving—that's a big deal. I received: underwear, boots, cap, pants, belt, clothing sack. And off you go! They gave me a dump truck. I moved concrete. There it was—and there it wasn't. We were young, unmarried. We didn't take any gas masks. There was one guy—he was older. He always wore his mask. But we didn't. The traffic guys didn't wear theirs. We were in the driver's cabin, but they were out in radioactive dust eight hours a day. Everyone got paid well: three times your salary plus vacation pay. We used it. We knew that vodka helped. It removed the stress. It's no wonder they gave people those 100 grams of vodka during the war. And then it was just like home: a drunk traffic cop fines a drunk driver.

Don't write about the wonders of Soviet heroism. They existed—and they really were wonders. But first there had to be incompetence, negligence, and only after those did you get wonders: covering the embrasure, throwing yourself in front of a machine gun. But that those orders should never have been given, that there shouldn't have been any need, no one writes about that. They flung us there, like sand onto the reactor. Every day they'd put out a new "Action Update": "men are working courageously and selflessly," "we will survive and triumph."

They gave me a medal and one thousand rubles.

*

At first there was disbelief, this sense that it was a game. But it was a real war, an atomic war. We had no idea—what's dangerous and what's not, what should we watch out for, and what ignore? No one knew.

It was a real evacuation, right to the train stations. What happened at the stations? We helped push kids through the windows of the train cars. We made the lines orderly—for tickets at the ticket window, for iodine at the pharmacy. In the lines people were swearing at one another and fighting. They broke the doors down on stores and stands. They broke the metal grates in the windows.

Then there were the people who came from somewhere else. They lived in clubs, schools, kindergartens. They walked around half-starving. Everyone's money ran out pretty fast. They bought up everything from the stores. I'll never forget the women who did the laundry. There were no washing machines, no one thought to bring those, so they washed by hand. All the women were elderly. Their hands were covered with boils and scabs. The laundry wasn't just dirty, it also had a few dozen roentgen. "Boys, have something to eat." "Boys, take a nap." "Boys, you're young, be careful." They felt sorry for us, they cried for us.

Are they still alive?

Every April 26, we get together, the guys who were there. We remember how it was. You were a soldier, at war, you were necessary. We forget the bad parts and remember that. We remember that they couldn't have made it without us. Our system, it's a military system, essentially, and it works great in emergencies. You're finally free there, and necessary. Freedom! And in those times the Russian shows how great he is. How unique. We'll never be Dutch or German. And we'll never have proper asphalt and manicured lawns. But there'll always be plenty of heroes.

*

They made the call, and I went. I had to! I was a member of the Party. Communists, march! That's how it was. I was a police officer—senior lieutenant. They promised me another "star." This was June of 1987. You were supposed to get a physical, but they just sent me without it. Someone, you know, got off, brought a note from his doctor that he had an ulcer, and I went in his place. It was urgent! [*Laughs.*] There were already jokes. Guy comes home from work, says to his wife, "They told me that tomorrow I either go to Chernobyl or hand in my Party card." "But you're not in the Party." "Right, so I'm wondering: how do I get a Party card by tomorrow morning?"

We went as soldiers, but at first they organized us into a masonry brigade. We built a pharmacy. Right away I felt weak and sleepy all the time. I told the doctor I was fine, it was just the heat. The cafeteria had meat, milk, sour cream from the kolkhoz, and we ate it all. The doctor didn't say anything. They'd make the food, he'd check with his book that everything was fine, but he never took any samples. We noticed that. That's how it was. We were desperate. Then the strawberries started coming, and there was honey everywhere.

The looters had already been there. We boarded up windows and doors. The stores were all looted, the grates on the windows broken in, flour and sugar on the floor, candy. Cans everywhere. One village got evacuated, and then five to ten kilometers over, the next village didn't. They brought all the stuff over from the evacuated village. That's how it was. We're guarding the place, and the former head of the kolkhoz arrives with some of the local people, they've already been resettled, they have new homes, but they've come back to collect the crops and sow new ones. They drove the straw out in bales. We found sewing machines and motorcycles in the bales. There was a barter system—they give you a bottle of homemade vodka, you give them permission to transport the television. We were selling and trading tractors and sowing machines. One bottle, or ten bottles. No one was interested in money. [*Laughs.*] It was like Communism. There was a tax for everything: a canister of gas—that's half a liter of vodka; an astrakhan fur coat—two liters; and motorcycles—variable. I spent six months there, that was the assignment. And then replacements came. We actually stayed a little longer, because the troops from the Baltic states refused to come. That's how it was. But I know people robbed the place, took out everything they could lift and carry. They transported the Zone back here. You can find it at the markets, the pawn shops, at people's dachas. The only thing that remained behind the wire was the land. And the graves. And our health. And our faith. Or my faith.

\*

We got to the place. Got our equipment. "Just an accident," the captain tells us. "Happened a long time ago. Three months. It's not dangerous anymore." "It's fine," says the sergeant. "Just wash your hands before you eat."

I measured the radiation. Once it got dark, these guys would pull up at our little station in cars and start giving us things: money, cigarettes, vodka. Just let us root around in the confiscated stuff. They'd pack their bags. Where'd they take it? Probably to Kiev and Minsk, to the second-hand markets. The stuff they left, we took care of it. Dresses, boots, chairs, harmonicas, sewing machines. We buried it in ditches—we called them "communal graves."

I got home, I'd go dancing. I'd meet a girl I liked and say, "Let's get to know one another."

"What for? You're a Chernobylite now. I'd be scared to have your kids."

\*

I have my own memories. My official post there was commander of the guard units. Something like the director of the apocalypse. [*Laughs.*] Yes. Write it down just like that.

I remember pulling over a car from Pripyat. The town is already evacuated, there aren't any people. "Documents, please."

They don't have documents. The back has a canvas cover. We lift it up, and I remember this clearly: twenty tea sets, a wall unit, an armchair, a television, rugs, bicycles.

So I write up a protocol.

I remember the empty villages where the pigs had gone crazy and were running around. The kolkhoz offices and clubs, these faded posters: "We'll give the motherland bread!" "Glory to the Soviet worker-peoples!" "The accomplishments of the people are immortal."

I remember the untended communal graves—a cracked headstone with men's names: Captain Borokin, Senior Lieutenant . . . And then these long columns, like poems—the names of privates. Around it, burdock, stinging-nettle, and goosefoot.

I remember this very nicely tended garden. The owner comes out of the house, sees us.

"Boys, don't yell. We already put in the forms—we'll be gone come spring."

"Then why are you turning over the soil in the garden?"

"But that's the autumn work."

I understand, but I have to write up a protocol . . .

*

My wife took the kid and left. That bitch! But I'm not going to hang myself, like Vanya Kotov. And I'm not going to throw myself out a seventh-floor window. That bitch! When I came back from there with a suitcase full of money, that was fine. We bought a car. That bitch lived with me fine. She wasn't afraid. [*Starts singing.*]

> *Even one thousand gamma rays*
> *Can't keep the Russian cock from having its days.*

Nice song. From there. Want to hear a joke? [*Launches right in.*] Guy comes home from the reactor. His wife asks the doctor, "What should I do with him?" "You should wash him, hug him, and put him out of commission." That bitch! She's afraid of me. She took the kid. [*Suddenly serious.*] The soldiers worked next to the reactor. I'd drive them there for their shifts and then back. I had a total-radiation-meter around my neck, just like everyone else. After their shifts, I'd pick them up and we'd go to the First Department—that was a classified department. They'd take our readings there, write something down on our cards, but the number of roentgen we got, that was a military secret. Those fuckers! Some time goes by and suddenly they say, "Stop. You can't take any more." That's all the medical information they give you. Even when I was leaving they didn't tell me how much I got. Fuckers! Now they're fighting for power. For cabinet portfolios. They have elections. You want another joke? After Chernobyl you can eat anything you want, but you have to bury your own shit in lead.

How are doctors going to work with us? We didn't bring any documents with us. They're still hiding them, or they've destroyed them because they were so classified. How do we help the doctors? If I had a certificate that said how much I got there? I'd show it to my bitch. I'll show her that we can survive anything and get married and have kids. The prayer of the Chernobyl liquidator: "Oh, Lord, since you've made it so that I can't, will you please also make it so I don't want to?" Oh, go fuck yourselves, all of you.

*

They made us sign a non-disclosure form. So I didn't say anything. Right after the army I became a second-group invalid.

I was twenty-two. I got a good dose. We lugged buckets of graphite from the reactor. That's ten thousand roentgen. We shoveled it with ordinary shovels, changing our masks up to thirty times a shift—people called them "muzzles." We poured the sarcophagus. It was a giant grave for one person, the senior operator, Valery Khodemchuk, who got caught under the ruins in the first minutes after the explosion. It's a twentieth-century pyramid. We still had three months left. Our unit got back, they didn't even give us a change of clothes. We walked around in the same pants, same boots, as we had at the reactor. Right up until they demobilized us.

And if they'd let me talk, who would I have talked to? I worked at a factory. My boss says: "Stop being sick or we'll have to cut you back." They did. I went to the director: "You have no right to do this, I'm a Chernobylite. I saved you. I protected you!" He says: "We didn't send you there."

At night I wake up from my mother saying, "Sonny, why aren't you saying anything? You're not asleep, you're lying there with your eyes open. And your light's on." I don't say anything. No one can speak to me in a way I can answer. In my own language. No one can understand where I've come back from. And I can't tell anyone.

*

I'm not afraid of death anymore. Of death itself. But I don't know how I'm going to die. My friend died. He got huge, fat, like a barrel. And my neighbor—he was also there, he worked a crane. He got black, like coal, and shrunk, so that he was wearing kids' clothes. I don't know how I'm going to die. I do know this: you don't last long with my diagnosis. But I'd like to feel it when it happens. Like if I got a bullet in the head.

I was in Afghanistan, too. It was easier there. They just shot you.

I clipped an article from the newspaper. It's about the operator Leonid Toptunov, he was the one on duty that night at the station and he pressed the red accident button a few minutes before the explosion. It didn't work. They took him to the hospital in Moscow. The doctors said, "In order to fix him, we'd need a whole other body." There was one tiny little non-radioactive spot on him, on his back. They buried him at the Mytinskaya Cemetery, like they did the others. They insulated the coffin with foil. And then they poured half a meter of concrete on it, with a lead cover. His father came. He's standing there, crying. People walk by: "That was your bastard son who blew it up!"

We're lonely. We're strangers here. They even bury us separately, not like they do other people. It's like we're aliens from outer space. I'd have been better off dying in Afghanistan. Honest, I get thoughts like that. In Afghanistan death was a normal thing. You could understand it there.

*

From above, from the helicopter, when I was flying near the reactor, I could see roes and wild boars. They were thin and sleepy, like they were moving in slow motion. They were eating the grass that grew there, and they didn't understand, they didn't understand that they should leave. That they should leave with the people.

Should I go or not go? Should I fly or not fly? I was a Communist—how could I not go?

Two paratroopers refused—their wives were young, they hadn't had any kids yet. But they were shamed and punished. Their careers were finished. And there was also a court of manhood, a court of honor! That was part of the attraction—he didn't go, so I will. Now I look at it differently. After nine operations and two heart attacks, I don't judge them, I understand them. They were young guys. But I would have gone anyway. That's definite. He couldn't, I will. That was manhood.

From above the amazing thing was the hardware: heavy helicopters, medium helicopters, the Mi-24, that's a fighting helicopter. What are you going to do with a Mi-24 at Chernobyl? Or with a fighter-plane, the Mi-2? The pilots, young guys, all of them fresh out of Afghanistan. Their feeling was they'd pretty much had enough, with Afghanistan, they'd fought enough. They're sitting in the forest near the reactor, catching roentgen. That was the order! They didn't need to send all those people there to get radiation. What for? They needed specialists, not a lot of human material. From above I saw a ruined building, a field of debris—and then an enormous number of little human shapes. There was a crane there, from East Germany, but it wasn't working—it made it to the reactor and then died. The robots died. Our robots, designed by Academic Lukachev for the exploration of Mars. And the Japanese robots—all their wiring was destroyed by the radiation, apparently. But there were soldiers in their rubber suits, their rubber gloves, running around . . .

Before we went back we were warned that in the interests of the State, it would be better not to go around telling people what we'd seen. But aside from us, no one knows what happened there. We didn't understand everything, but we saw it all.

# PART TWO

# THE LAND OF THE LIVING

### MONOLOGUE ABOUT OLD PROPHECIES

My little daughter—she's different. She's not like the others. She's going to grow up and ask me: "Why aren't I like the others?"

When she was born, she wasn't a baby, she was a little sack, sewed up everywhere, not a single opening, just the eyes. The medical card says: "Girl, born with multiple complex pathologies: aplasia of the anus, aplasia of the vagina, aplasia of the left kidney." That's how it sounds in medical talk, but more simply: no pee-pee, no butt, one kidney. On the second day I watched her get operated on, on the second day of her life. She opened her eyes and smiled, and I thought that she was about to start crying. But, God, she smiled!

The ones like her don't live, they die right away. But she didn't die, because I loved her.

In four years she's had four operations. She's the only child in Belarus to have survived being born with such complex pathologies. I love her so much. [*Stops.*] I won't be able to give birth again. I wouldn't dare. I came back from the maternity ward, my husband would start kissing me at night, I would lie there and tremble: we can't, it's a sin, I'm scared. I heard the doctors talking: "That girl wasn't born in a shirt, she was born in a suit of armor. If we showed it on television not a single mother would give birth." That was about our daughter. How are we supposed to love each other after that?

I went to church and told the minister. He said I needed to pray for my sins. But no one in my family has ever killed anyone. What am I guilty of? First they wanted to evacuate our village, and then they crossed it off their lists—the government didn't have enough money. And right around then I fell in love. I got married. I didn't know that we weren't allowed to love here. Many years ago, my grandmother read in the Bible that there will be a time when everything is thriving, everything blossoming and fruitful, and there will be many fish in the rivers and animals in the forest, but man won't be able to use any of it. And he won't be able to propagate himself in his likeness, to continue his line. I listened to the old prophecies like they were scary fairy tales. I didn't believe them.

Tell everyone about my daughter. Write it down. She's four years old and she can sing, dance, she knows poetry by heart. Her mental development is normal, she isn't any different from the other kids, only her games are different. She doesn't play "store," or "school"—she plays "hospital." She gives her dolls shots, takes their temperature, puts them on IV. If a doll dies, she covers it with a white sheet. We've been living in the hospital with her for four years, we can't leave her there alone, and she doesn't even know that you're supposed to live at home. When we go home for a month or two, she asks me, "When are we going back to the hospital?" That's where her friends are, that's where they're growing up.

They made an anus for her. And they're forming a vagina. After the last operation her urinary functioning completely broke down, and they were unable to insert a catheter—they'll need more operations for that. But from here on out they've

advised us to seek medical help abroad. Where are we going to
get tens of thousands of dollars if my husband makes 120 dollars
a month? One professor told us quietly: "With her pathologies,
your child is of great interest to science. You should write to
hospitals in other countries. They should be interested." So I
write. [*Tries not to cry.*] I write that every half hour we have to
squeeze out her urine manually, it comes out through artificial
openings in the area of her vagina. Where else is there a child
in the world who has to have her urine squeezed out of her
every half hour? And how much longer can it go on? No one
knows the effect of small doses of radiation on the organism
of a child. Take my girl, even if it's to experiment. I don't want
her to die. I'm all right with her becoming a lab frog, a lab
rabbit, just as long as she lives. [*Cries.*] I've written dozens of
letters. Oh, God!

She doesn't understand yet, but someday she'll ask us: why
isn't she like everyone else? Why can't she love a man? Why
can't she have babies? Why won't what happens to butterflies
ever happen to her? What happens to birds? To everyone but
her? I wanted—I should have been able to prove—so that—I
wanted to get papers—so that she'd know—when she grew
up—it wasn't our fault, my husband and I, it wasn't our love
that was at fault. [*Tries again not to cry.*] I fought for four
years—with the doctors, the bureaucrats—I knocked on the
doors of important people. It took me four years to finally
get a paper from the doctors that confirmed the connection
between ionized radiation (in small doses) and her terrible
condition. They refused me for four years, they kept telling
me: "Your child is a victim of a congenital handicap." What
congenital handicap? She's a victim of Chernobyl! I studied
my family tree—nothing like this has ever happened in our
family. Everyone lived until they were eighty or ninety. My

grandfather lived until he was 94. The doctors said: "We have instructions. We are supposed to call incidentes of this type general sicknesses. In twenty or thirty years, when we have a database about Chernobyl, we'll begin connecting these to ionized radiation. But for the moment science doesn't know enough about it." But I can't wait twenty or thirty years. I wanted to sue them. Sue the government. They called me crazy, laughed at me, like, There were children like these in ancient Greece, too. One bureaucrat yelled at me: "You want Chernobyl privileges! Chernobyl victim funds!" How I didn't lose consciousness in his office, I'll never know.

There was one thing they didn't understand—didn't want to understand—I needed to know that it wasn't our fault. It wasn't our love. [*Breaks down. Cries.*] This girl is growing up—she's still a girl—I don't want you to print our name—even our neighbors—even other people on our floor don't know. I'll put a dress on her, and a handkerchief, and they say, "Your Katya is so pretty." Meanwhile I give pregnant women the strangest looks. I don't look at them, I kind of glance at them real quick. I have all these mixed feelings: surprise and horror, jealousy and joy, even this feeling of vengeance. One time I caught myself thinking that I look the same way at the neighbors' pregnant dog—at the bird in its nest . . .

My girl . . .

*Larisa Z., mother*

## MONOLOGUE ABOUT A MOONLIT LANDSCAPE

I suddenly started wondering about what's better—to remember or to forget? I asked my friends. Some have forgotten, others don't want to remember, because we can't change anything anyway, we can't even leave here.

Here's what I remember. In the first days after the accident, all the books at the library about radiation, about Hiroshima and Nagasaki, even about X-rays, disappeared. Some people said it was an order from above, so people wouldn't panic. There was even a joke that if Chernobyl had blown up near the Papuans, the whole world would be frightened, but not the Papuans. There were no medical bulletins, no information. Those who could got potassium iodide (you couldn't get it at the pharmacy in our town, you had to really know someone). Some people took a whole bunch of these tablets and washed them down with liquor. Then they had to get their stomachs pumped at the hospital. Then we discovered a sign, which all of us followed: as long as there were sparrows and pigeons in town, humans could live there, too. I was in a taxi one time, the driver couldn't understand why the birds were all crashing into his window, like they were blind. They'd gone crazy, or like they were committing suicide.

I remember coming back one time from a business trip. There was a moonlit landscape. On both sides of the road, to the very horizon, stretched these fields covered in white dolomite. The poisoned topsoil had been removed and buried, and in its place they brought white dolomite sand. It was like not-earth. This vision tortured me for a long time and I tried to write a story. I imagined what would be here in a hundred years: a person, or something else, would be galloping along on all fours, throwing out its long back legs, knees bent. At night it can see with a third eye, and its only ear, on the crown of its head, can even hear how ants run. Ants are the only thing left, everything else in heaven and earth has died.

I sent the story to a journal. They wrote back saying that this wasn't a work of literature, but the description of a nightmare. Of course I lacked the talent. But there was another reason they didn't take it, I think.

I've wondered why everyone was silent about Chernobyl, why our writers weren't writing much about it—they write about the war, or the camps, but here they're silent. Why? Do you think it's an accident? If we'd beaten Chernobyl, people would talk about it and write about it more. Or if we'd understood Chernobyl. But we don't know how to capture any meaning from it. We're not capable of it. We can't place it in our human experience or our human time-frame.

So what's better, to remember or to forget?

*Yevgeniy Aleksandrovich Brovkin,*
*instructor at Gomel State University*


## MONOLOGUE ABOUT A MAN WHOSE TOOTH WAS HURTING WHEN HE SAW CHRIST FALL

I was thinking about something else then. You'll find this strange, but I was splitting up with my wife.

They came suddenly, gave me a notice, and said, There's a car waiting downstairs. It was like 1937. They came at night to take you out of your warm bed. Then that stopped working: people's wives would refuse to answer the door, or they'd lie, say their husbands were away on business, or vacation, or at the dacha with their parents. The soldiers would try to give them the notice, the wives would refuse to take it. So they started grabbing people at work, on the street, during a lunch break at the factory cafeteria. It was just like 1937.

But I was almost crazy by then. My wife had cheated on me, everything else didn't matter. I got in their car. The guys who came for me were in street clothes, but they had a military bearing, and they walked on both sides of me, they were clearly worried I'd run off. When I got in the car, I remembered for some reason the American astronauts who'd flown to the moon, and one of them later became a priest, and the other apparently went crazy. I read that they thought they'd seen cities, some kind of human remnants there. I remembered some lines from the papers: our nuclear stations are absolutely safe, we could build one on Red Square, they're safer than samovars. They're like stars and we'll "light" the whole earth with them. But my wife had left me, and I could only think about that. I tried to kill myself a few times. We went to the same kindergarten, the same school, the same college. [*Silent. Smokes.*]

I told you. There's nothing heroic here, nothing for the writer's pen. I had thoughts like, It's not wartime, why should I have to risk myself while someone else is sleeping with my wife? Why me again, and not him? To be honest, I didn't see any heroes there. I saw nutcases, who didn't care about their own lives, and I had enough craziness myself, but it wasn't necessary. I have medals and awards—but that's because I wasn't afraid of dying. I didn't care! It was even something of an out. They'd have buried me with honors, and the government would have paid for it.

You immediately found yourself in this fantastic world, where the apocalypse met the stone age. And for me it was sharper, barer. We lived in the forest, in tents, twenty kilometers from the reactor, like partisans. Partisans are the people who get called up for military training. We were between twenty-five and forty, some of us had university degrees, or vocational-technical degrees. I'm a history teacher, for example. Instead of machine guns they gave us shovels. We buried trash heaps and gardens. The women in the villages watched us and crossed themselves. We had gloves, respirators, and surgical robes. The sun beat down on us. We showed up in their yards like demons. They didn't understand why we had to bury their gardens, rip up their garlic and cabbage when it looked like ordinary garlic and ordinary cabbage. The old women would cross themselves and say, "Boys, what is this—is it the end of the world?"

In the house the stove's on, the lard is frying. You put a dosimeter to it, and you find it's not a stove, it's a little nuclear reactor. "Boys," the men say, "have a seat at the table." They want to be friendly. We say no. They say, "Come, we'll drink a hundred grams. Have a seat. Tell us what's going on." What do we tell them? At the reactor the firefighters were stomping on burning fuel, and it was glowing, but they didn't know what it was. What's to know? We go in units, and each unit has one dosimeter. Different places have different levels of radiation. One of us is working where there are two roentgen, but another guy is working where there are ten. On the one hand, we have no rights, like prisoners, and on the other hand, we're frightened. But I wasn't frightened. I was watching everything from the side.

A group of scientists flew in on a helicopter. In special rubber suits, tall boots, protective goggles. Like they were going to the moon. This old woman comes up to one of them and says, "Who are you?" "I'm a scientist." "Oh, a scientist. Look how he's dressed up! Look at that mask! And what about us?" And she goes after him with a stick. I've thought a few times that someday they're going to start hunting the scientists the way they used to hunt the doctors and drown them in the Middle Ages.

I saw a man who watched his house get buried. [*Stops.*] We buried houses, wells, trees. We buried the earth. We'd cut things down, roll them up into big plastic sheets . . . I told you, nothing heroic here.

One time we're coming back late at night—we worked twelve-hour shifts, without any holidays, so the only time we could rest was at night. So we're in the APC and we see a person walking through this abandoned village. We come a little closer and it's a young guy with a rug on his back.

There's a Zhiguli nearby. We stop, have a look: the trunk is stuffed with televisions and telephones. The APC turns around and wham: the Zhiguli just collapses, like a soda can. No one says a word.

We buried the forest. We sawed the trees into meter-and-a-half pieces and packed them in cellophane and threw them into graves. I couldn't sleep at night. I'd close my eyes and see something black moving, turning over—as if it were alive—live tracts of land—with bugs, spiders, worms—I didn't know any of them, what they were called, just bugs, spiders, ants. And they were small and big, yellow and black, all different colors. One of the poets says somewhere that animals are a different people. I killed them by the ten, by the hundred, thousand, not even knowing what they were called. I destroyed their houses, their secrets. And buried them. Buried them.

Leonid Andreev, whom I love very much, has this parable about Lazarus, who looked into the abyss. And now he's alien, he'll never be the same as other people, even though Christ resurrected him.

Maybe that's enough? I know you're curious, people who weren't there are always curious. But it was still a world of people, the same one. It's impossible to live constantly in fear, a person can't do it, so a little time goes by and normal human life resumes. [*Continues.*] The men drank vodka. They played cards, tried to get girls, had kids. They talked a lot about money. But it wasn't for money that we went there. Or most people didn't. Men worked because you have to work. They told us to work. You don't ask questions. Some hoped for better careers out of it. Some robbed and stole. People hoped for the privileges that had been promised: an apartment without waiting and moving out of the barracks, getting their kid into a kindergarten, a car. One guy got scared, refused to leave the tent, slept in his plastic suit. Coward! He got kicked out of the Party. He'd yell, "I want to live!" There were all kinds of people. I met women there who'd volunteered to come, who'd demanded to come. They were told no, we need chauffers, plumbers, firemen, but they came anyway. All kinds of people. Thousands of volunteers guarding the storehouses at night. There were student units, and wire transfers to the fund for victims. Hundreds of people who donated blood and bone marrow.

And at the same time you could buy anything for a bottle of vodka. A medal, or sick leave. One kolkhoz chairman would bring a case of vodka to the radiation specialists so they'd cross his village off the lists for evacuation; another would bring the same case so that they'd put his village *on* the list—he's already been promised a three-room apartment in Minsk. No one checked the radiation reports. It was just your average Russian chaos. That's how we live. Some things were written off and sold. On the one hand, it's disgusting, and on the other hand—why don't you all go fuck yourselves?

They sent students. They pulled the goose-foot out in the fields. Collected straw. A few couples were really young, a husband and wife. They were still walking around holding hands. That was impossible to watch. And the place was so beautiful! Really incredible. The horror was more horrible because it was so pretty. And people had to leave here. They had to run away, like evildoers, like criminals.

Every day they brought the paper. I'd just read the headlines: "Chernobyl—A Place of Achievement." "The Reactor Has Been Defeated!" "Life Goes On." We had political officers, they'd hold political discussions with us. We were told that we had to win. Against whom? The atom? Physics? The universe? Victory is not an event for us, but a process. Life is a struggle. An overcoming. That's why we have this love of floods and fires and other catastrophes. We need an opportunity to demonstrate our "courage and heroism."

Our political officer read notices in the paper about our "high political consciousness and meticulous organization," about the fact that just four days after the catastrophe the red flag was already flying over the fourth reactor. It blazed forth. In a month the radiation had devoured it. So they put up another flag. And in another month they put up another one. I tried to imagine how the soldiers felt going up on the roof to replace that flag. These were suicide missions. What would you call this? Soviet paganism? Live sacrifice? But the thing is, if they'd given me the flag then, and told me to climb up there, I would have. Why? I can't say. I wasn't afraid to die, then. My wife didn't even send a letter. In six months, not a single letter. [*Stops.*] Want to hear a joke? This prisoner escapes from jail, and runs to the thirty-kilometer zone at Chernobyl. They catch him, bring him to the dosimeters. He's "glowing" so much, they can't possibly put him back in prison, can't take him to the hospital, can't put him around people.

Why aren't you laughing?

[*Laughs.*]

When I got there, the birds were in their nests, and when I left the apples were lying on the snow. We didn't get a chance to bury all of them. We buried earth in the earth. With the bugs, spiders, leeches. With that separate people. That world. That's my most powerful impression of that place—those bugs.

I haven't told you anything, really. Just snippets. The same Leonid Andreev has a parable about a man who lived in Jerusalem, past whose house Christ was taken, and he saw and heard everything, but his tooth hurt. He watched Christ fall while carrying the cross, watched him fall and cry out. He saw all of this, but his tooth hurt, so he didn't run outside. Two days later, when his tooth stopped hurting, people told him that Christ had risen, and he thought: "I could have been a witness to it. But my tooth hurt."

Is that how it always is? My father defended Moscow in 1942. He only learned that he'd been part of a great event many years later, from books and films. His own memory of it was: "I sat in a trench. Shot my rifle. Got buried by an explosion. They dug me out half-alive." That's it.

And back then, my wife left me.

*Arkady Filin, liquidator*

THREE MONOLOGUES ABOUT A SINGLE BULLET

*Speakers: Viktor Iosifovich Verzhikovskiy, chairman of the Khoyniki Society of Volunteer Hunters and Fishermen, and two hunters, Andrei and Vladimir, who did not want their full names used.*

The first time I killed a fox, I was a kid. The next time I killed a doe, and then I swore never to kill another one. They have such expressive eyes.

It's us, people, who understand things. Animals just live. So do birds.

During the fall the wild goat is very sharp. If there's the slightest wind from humans, that's it, she won't let you near. Whereas the fox is very clever . . .

They say there used to be this guy, he'd walk around. If he got drunk, he'd start reading everyone lectures. He'd studied philosophy at the university, then he'd been in prison. You meet someone in the Zone, they'll never tell you the truth about themselves. Or very rarely. But this one was intelligent. "Chernobyl," he'd say, "happened so that philosophers could be made." He called animals "walking ashes," and people, "talking earth." The earth talked because we eat earth, that is, we are built from earth.

The Zone pulls you in. You miss it, I tell you. Once you've been there, you'll miss it.

All right, boys, but let's do this in order.

Right, right, chairman. You tell it, we'll smoke a bit.

So, it was like this. They call me into the regional executive. "Listen, chairman hunter. There are still many household pets in the Zone—cats, dogs. In order to avoid an epidemic, we need to exterminate them. Go to it!" On the next day I called everyone together, all the hunters. I explained the situation. No one wants to go because they haven't given us any protective gear. I asked the civil defense people, and they didn't have anything. Not a single respirator. I ended up having to go to the cement factory and get masks from them. Just a thin little bit of film, to protect against cement dust. But no respirators.

We met soldiers there. They had masks, gloves, they had armored personnel carriers. And we're in shirts, and a handkerchief over our noses. And we came home in those shirts and those boots, to our families.

I got together two brigades, twenty men each. Each brigade had a veterinarian with it and someone from the epidemic center. We also had a tractor with a scooper and a dump truck. It's too bad they didn't give us any protection, didn't think about the people.

On the other hand they gave us rewards—thirty rubles apiece. A bottle of vodka cost three rubles then. And when we got deactivated, people came up with these recipes: a spoonful of goose shit onto a bottle of vodka. Drink this for two days. So that, you know—so that you could, as a man . . . We had that little chastushka, remember? There were tons of them. "The Zaporozhets can't keep it up; the man from Kiev can't get it up. If you want to be a father, wrap your nuts in a lead nut-warmer." Ha-ha.

We rode around the Zone for two months. Half the villages in our region were evacuated, dozens of them: Babchin, Tulgovichi . . . The first time we came, the dogs were running around near their houses, guarding them. Waiting for the people to come back. They were happy to see us, they ran toward our voices. We shot them in the houses, and the barns, in the yards. We'd drag them out onto the street and load them onto the dump truck. It wasn't very nice. They couldn't understand: why are we killing them? They were easy to kill. They were household pets. They didn't fear guns or people. They ran toward our voices.

There was this turtle crawling . . . God! Past an empty house. There were aquariums in the houses, with fish in them.

We didn't kill the turtles. If you ran over a turtle with your jeep, the shell held up. It didn't crack. Of course we only did this when we were drunk. There were cages open in the yards. Rabbits running around. The otters were shut in, we let them out, if there was water nearby, a lake or a river, they'd swim away.

Everything was abandoned. For a time. Because what was the order? "Three days." They'd trick the little kids: "We're going to the circus." They'd cry. And people thought they'd come back. I'll tell you, it was a war zone. The cats looked people in the eye, the dogs howled, trying to get on the buses. The mutts and the shepherds both. The soldiers pushed them out. Kicked them. They ran a long way after the cars. An evacuation—it's a terrible thing.

So here's how it is. The Japanese had Hiroshima, and now they're ahead of everyone. They're at the top. So that means . . .

It's a chance to shoot at something, and it's moving and alive. It's an instinct. It's fun. We'd drink, and then go. We were getting paid at work, which is fair considering what we were doing. And then thirty rubles—back then—under the Communists—you could . . .

It was like this. At first the houses were sealed. We didn't tear the seals off. If you could see a cat through the window, how were you going to get at it? We didn't touch them. Then the robbers started coming through, breaking down the doors, windows, the window-guards. They stole everything. First they took the record players and televisions, the fur clothing. Then they took everything else. The floor would be littered with just aluminum spoons. And then the dogs that were still alive would move into the house. You come in, the dog comes at you. At this point they'd stopped trusting humans. I came in one time, and there's a bitch lying in the middle of the room with her little puppies around her. Did I feel sorry for her? Sure, it wasn't pleasant. But I compared it. In effect it was like we were at war, we were punishers. It was the same kind of scheme. A military operation. We also came, surrounded the village, and the dogs, once they hear the first shot, run away. Into the forest. The cats are smarter, and it's easier for them to hide. One cat got into a clay pot. I shook him out of there. We pulled them out from beneath stoves. You got an unpleasant feeling. You walk into the house, and the cat zips by like a bullet past your foot, you run after it with your rifle. They're thin, dirty. Their fur's all in clumps. At first there were a lot of eggs, the chickens were still there hatching them. So the dogs and cats ate the eggs, and when the eggs were finished they ate the chickens. And the foxes also ate the chickens, the foxes were already living in the villages with the dogs. So then there were no more chickens and the dogs started eating the cats. There were times we'd find some pigs in a barn, and we'd let them out, and then in the cellars there's all sorts of things: cucumbers, tomatoes. We open them up and throw them in the trough. We didn't kill the pigs.

There was this old lady, in one of the villages, she shut herself up in her house. She had five cats and three dogs. She wouldn't give them up. Cursed us. We took them by force. We left her one cat and one dog. She cursed. She called at us: "Bandits! Jailers!"

The empty villages, just the stoves. Then there's Khatyni. In the middle of Khatyni there are these two old ladies. And they're not afraid. Anyone else would have gone crazy.

Yeah, ha. "Next to the hill you're on your tractor, across the way there's the reactor. If the Swedes hadn't've told, we'd be on the tractor, getting old." Ha-ha.

So it was like this. The smells—I couldn't understand where this smell was coming from in the village. Six kilometers from the reactor. The village of Masaly. It was like roentgen central. It smelled of iodine. Some kind of sourness. You had to shoot them point blank. This bitch is on the floor with her pups, she jumps right at me. I shoot her quick. The puppies lick their paws, fawning, playing around. I had to shoot them point blank. One dog—he was a little black poodle. I still feel sorry for him. We loaded a whole dump truck with them, even filled the top. We drive them over to our "cemetery." To be honest it was just a deep hole in the ground, even though you're supposed to dig it in such a way that you can't reach any ground water, and you're supposed to insulate it with cellophane. You're supposed to find an elevated area. But of course those instructions were violated everywhere. There wasn't any cellophane, and we didn't spend a lot of time looking for the right spot. If they weren't dead, if they were just wounded, they'd start howling, crying. We're dumping them from the dump truck into the hole, and this one little poodle is trying to climb back out. No one has any bullets left. There's nothing to finish him with. Not a single bullet. We pushed him back into the hole and just buried him like that. I still feel sorry for him.

But there were a lot fewer cats than dogs. Maybe they left after the people? Or they hid? It was a little household poodle, a spoiled poodle.

It's better to kill from far away, so your eyes don't meet.

You have to learn to shoot accurately, so you don't have to finish them off later.

It's us, people, who understand things, but they just live. "Walking ashes."

Horses—when you took them to be shot, they'd cry.

And I'll add this—any living creature has a soul, even insects. This wounded doe—she's lying there. She wants you to feel sorry for her, but instead you finish her off. At the last moment she has an understanding, almost human look. She hates you. Or it's a plea: I also want to live! I want to live!

Learn to shoot, I tell you! Beating them is much worse than killing them. Hunting is a sport, a kind of sport. For some reason no one bothers the fishermen, but everyone bothers the hunters. It's unfair!

Hunting and war—these are the main activities for a man. For a real man.

I couldn't tell my son about it. He's a kid. Where was I? What was I doing? He still thinks his father was over there defending someone or something. That he was at his battle station! They showed it on television: military equipment, lots of soldiers. There were a lot of soldiers. My son asks me: "Papa, you were like a soldier?"

This cameraman came with us from the television. Remember? He cried. He was a man but he cried. He kept wanting to see a three-headed boar.

Yeah, ha. The fox sees how a gingerbread man is rolling through the forest. "Gingerbread man, where are you rolling to?" "I'm not a gingerbread man, I'm a hedgehog from Chernobyl." Ha-ha. Like they say, let's put the peaceful atom into every home!

I'll tell you, every person dies just like an animal. I saw this many times in Afghanistan. I, myself, I was wounded there in the stomach, and I was lying in the sun. The heat was unbearable. I was thirsty! "Well," I thought, "I'm going to die here, like a dog." I'll tell you, the blood flows the same way, just like theirs does, and the pain is the same.

The police officer that was with us—he went crazy. He felt sorry for the Siamese cats, they were so expensive on the market, he said. They were pretty. And he was a man . . .

A cow is walking with her calf. We don't shoot. And we don't shoot the horses either. They were afraid of wolves, but not of people. But a horse can defend itself. The wolves got the cows first. It was the law of the jungle.

They shipped the cattle to Russia from Belarus and sold it. Meanwhile the heifers were leukemic. But they gave discounts on those.

I feel worst for the old men. They'd come up to our cars: "Son, will you have a look at my house?" Giving me the keys. "Could you grab my suit? And my hat." Giving me a few coins. "How is my dog doing?" The dog's been shot, the house has been looted. And they'll never return there. How do you tell them? I didn't take the keys. I didn't want to trick anyone. Others took them. "Where'd you put the vodka? Where'd you hide it?" And the old man would tell them. They'd find whole milk cans full of the stuff.

They asked us to kill a wild boar for a wedding. It was a request. The liver melted in your hands, but they wanted it anyway, for the wedding. For the christening.

We shoot for science, too. One time we shot two rabbits, two foxes, two wild goats. They're all sick, but we still tenderize them and eat. At first we were afraid to, but now we're used to it. You have to eat something, and we can't all move to the moon, to another planet.

Someone bought a fox-fur hat at the market, and he went bald. An Armenian bought a cheap machine gun from a guy from the Zone—he died. People frighten one another.

As for me, nothing happened there to my soul or my mind. That's all a lot of nonsense.

I talked to a driver who was transporting homes out of there. He was just driving them out. Of course these aren't houses or schools or kindergartens anymore, they're just numbered objects of deactivation. But they bring them out! I met him in the bathhouse, or maybe at a beer stand. I don't remember exactly, but he was telling me: they bring the truck, in three hours they take apart the house and put it in there, and then at the edge of the Zone someone meets them. They just tear it apart. The Zone has been sold for dacha parts. The driver gets some money and they feed him and get him drunk.

Some of us are predators—hunter-predators. Others like to just walk in the forest, go after the small game. The birds.

I'll tell you: So many people suffered, and no one ever answered for it. They put away the director of the station, but then they let him out again. In that system, it was hard to say who was guilty. They were trying something out there. I read in the paper that they were developing military plutonium. For atom bombs. That's why it blew up. But if that's why it happened, then why here? Why at Chernobyl? Why not in France or in Germany?

This one thing stuck in my memory. That one thing. No one had a single bullet, there was nothing to shoot that little poodle with. Twenty guys. Not a single bullet at the end of the day. Not a single one.

## MONOLOGUE ABOUT HOW WE CAN'T LIVE WITH OUT CHEKHOV AND TOLSTOY

What do I pray for? Ask me: what do I pray for? I don't pray in church. I pray to myself. I want to love! I do love. I pray for my love! But for me—[*Stops short. I can see she doesn't want to talk.*] Am I supposed to remember? Maybe I should push it away instead, just in case? I never read such books. I never saw such movies. At the movies I saw the war. My grandmother and grandfather remember that they never had a childhood, they had the war. Their childhood is the war, and mine is Chernobyl. That's where I'm from. You're a writer, but no book has helped me to understand. And the theater hasn't, and the movies haven't. I understand it without them, though. By myself. We all live through it by ourselves, we don't know what else to do. I can't understand it with my mind. My mother especially has felt confused. She teaches Russian literature, and she always taught me to live with books. But there are no books about this. She became confused. She doesn't know how to do without books. Without Chekhov and Tolstoy. Am I supposed to remember? I want to remember, and also I don't want to. [*Either she's listening to herself, or arguing with herself.*] If scientists don't know anything, if writers don't know anything, then we'll help them with our own lives and our deaths. That's what my mom thinks. But I don't want to think about this, I want to be happy. Why can't I be happy?

We lived in Pripyat, near the nuclear station, that's where I was born and grew up. In a big pre-fab building, on the fifth floor. The windows looked out onto the station. On April 26—there were two days—those were the last two days in our town. Now it's not there anymore. What's left there isn't our town. That day a neighbor was sitting on the balcony, watching the fire through binoculars. Whereas we—the girls and boys—we raced to the station on our bikes, and those who didn't have bikes were jealous. No one yelled at us not to go. No one! Not our parents, not our teachers. By lunch time there weren't any fishermen at the river, they'd come back black, you can't get that black in a month at Sochi. It was a nuclear tan! The smoke over the station wasn't black or yellow, it was blue. But no one yelled at us. People were used to military dangers: an explosion over here, an explosion over there. Whereas here you had an ordinary fire, being put out by ordinary firemen. The boys were joking around: "Get in a row at the cemetery, whoever's tallest dies first." I was little. I don't remember the fear, but I remember lots of weird things. My friend was telling me that she and her mother spent the night burying their money and gold things, and were worried they'd forget the spot. My grandmother, when she'd retired, had been given a samovar from Tula, and for some reason the thing she worried about most was the samovar, and also about Grandpa's medals. And about the old Singer sewing machine. We were "evacuated." My father brought that word home from work. It was like in the war books. We were already on the bus when my father remembered he'd left something. He runs home, comes back with two of his new shirts still on their hangers. That was strange. The soldiers were sort of like aliens, they walk through the streets in their protective gear and masks. "What's going to happen to us?" people were asking them. "Why are you asking us?" they'd snap back. "The white Volgas are over there, that's where the bosses are, ask them."

We're riding on the bus, the sky is blue as blue. Where are we going? We have Easter cakes and colored eggs in our bags and baskets. If this is war, it's not how I imagined it from the books I'd read. There should have been explosions over here, over there, bombing. We were moving slowly, the livestock was in the way. People were chasing cows and horses down the roads. It smelled of dust and milk. The drivers were cursing and yelling at the shepherds: "Why are you on the road with those, you this-and-that?? You're kicking up radioactive dust! Why don't you take them through the fields?" And those cursed back that it'd be a shame to trample all the rye and grass. No one thought we'd never come back. Nothing like this had ever happened. My head was spinning a little and my throat tickled. The old women weren't crying, but the young ones were. My mother was crying.

We got to Minsk. But we had to buy our seats on the train at triple the usual price. The conductor brought everyone tea, but to us she says, "Let me have your cups." We didn't get it right away—did they run out of cups? No! They were afraid of us. "Where are you from?" "Chernobyl." And then they shy off. In a month my parents were allowed to go to the apartment. They got a warm blanket, my fall coat and the collected letters of Chekhov, my mom's favorite. Grandma—our grandma—she couldn't understand why they didn't take the cans of strawberry jam she'd made—they were in cans, after all, they were sealed up. They found a "stain" on the blanket. My mother washed it, vacuumed it, nothing helped. They gave it to the dry-cleaners, it turned out the spot "glowed." They cut it out with their scissors. It was the same blanket, and my same coat, but I couldn't sleep under the blanket anymore, or wear that coat. It wasn't that I was afraid of those things—I hated them! Those things could have killed me! I felt this animosity—I don't really understand it myself.

Everyone was talking about the accident: at home, in school, on the bus, in the street. People compared it to Hiroshima. But no one believed it. How can you believe in something incomprehensible? No matter how hard you try, it still doesn't make sense. I remember—we're leaving, the sky is blue as blue. And Grandma—she couldn't get used to the new place. She missed our old home. Just before she died she said, "I want some sorrel!" We weren't allowed to eat that for several years, it was the thing that absorbed the most radiation.

We buried her in her old village of Dubrovniki. It was in the Zone, so there was barbed wire and soldiers with machine guns guarding it. They only let the adults through—my parents and relatives. But they wouldn't let me. "Kids aren't allowed." I understood then that I would never be able to visit my grandmother. I understood. Where can you read about that? Where has that ever happened? My mom admitted: "You know, I hate flowers and trees." She became afraid of herself. At the cemetery, on the grass, they put down a table-cloth and placed some food and vodka on it, for the wake. The soldiers brought over the dosimeter and threw everything out. The grass, the flowers, everything was "clicking." Where did we take our grandma?

I'm afraid. I'm afraid to love. I have a fiancé, we already registered at the house of deeds. Have you ever heard of the Hibakusha of Hiroshima? The ones who survived after the bomb? They can only marry each other. No one writes about it here, no one talks about it, but we exist. The Chernobyl Hibakusha. He brought me home to his mom, she's a very nice mom. She works at a factory as an economist, and she's very active, she goes to all the anti-Communist meetings. So this very nice mom, when she found out that I'm from a Chernobyl family, a refugee, asked: "But, my dear, will you be able to have children?" And we've already registered! He pleads with me: "I'll leave home. We'll rent an apartment." But all I can hear is: "My dear, for some people it's a sin to give birth." It's a sin to love.

Before him I had another boyfriend. He was an artist. We also wanted to get married. Everything was fine until this one thing happened. I came into his studio and heard him yelling into the phone: "You're lucky! You have no idea how lucky you are!" He's usually so calm, even phlegmatic, not a single exclamation point in his speech. And then this! So what is it? Turns out his friend lives in a student dormitory, and he looked into the next room, and there's a girl hanging there. She strung herself up with some panty hose. He takes her down. And my boyfriend was just beside himself, shivering: "You have no idea what he's seen! What he's just been through! He carried her in his arms—he touched her face. She had white foam on her lips. Maybe if we hurry we can make it." He didn't mention the dead girl, didn't feel sorry for her for a second. He just wanted to see it and remember it. So he could draw it later on. And I started remembering how he used to ask me what color the fire at the station was, and whether I'd seen cats and dogs that had been shot, were they lying on the street? Were people crying? Did I see how they died? After that . . . I couldn't be with him anymore. I couldn't answer him. [*After a pause.*] I don't know if I'd want to meet with you again. I think you look at me the same way he did. Just observing me and remembering. Like there's an experiment going on. I can't rid myself of that feeling. I'll never rid myself of it.

Do you know that it can be a sin to give birth? I'd never heard those words before.

*Katya P.*

## MONOLOGUE ABOUT WAR MOVIES

This is my secret. No one else knows about this. I've only talked about this with one friend.

I'm a cameraman. I went there with everything we'd been taught: at war, you become a real writer. *Farewell to Arms* was my favorite book. So I got there. People are digging in their gardens, there are tractors and seed drills in the fields. What do I film? Nothing's blowing up.

My first shoot was in an agricultural club. They put a television on the stage and gathered everyone together. They listened to Gorbachev—everything's fine, everything's under control. In the village where we were shooting they were doing a "de-activation." They were washing roofs. But how do you wash an old lady's roof if it leaks? As for the soil, you had to cut off the entire fertile layer of it. After that there's yellow sand. One old lady was following orders and throwing the earth out, but then scraping the manure off to use later. It's too bad I didn't shoot that.

Everywhere you went, people would say, "Ah, movie people. Hold on, we'll find you some heroes." And they'd produce an old man and his grandson who spent two days chasing cows off from right near Chernobyl. After the shoot the livestock specialist calls me over to a giant pit, where they're burying the cows with a bulldozer. But it didn't even occur to me to shoot that. I turned my back on the pit and shot the scene in the great tradition of our patriotic documentaries: the bulldozer drivers are reading *Pravda*, the headline in huge block letters: "The nation will not abandon those in trouble!" I even got lucky: I look and there's a stork landing in a field. A symbol! No matter what catastrophes befall us, we will triumph! Life goes on!

The country roads. Dust. I already knew this wasn't just dust, but radioactive dust. I hid the camera so as to protect the optics. It was a very dry May. I don't know how much we swallowed. A week in, my lymph nodes swelled up. But we were conserving film like it was ammunition, because the first secretary of the Central Committee of Belarus, Slyunkov, was supposed to come. No one would tell us where exactly he'd be arriving, but we figured it out. I was driving down the road one day, the dust was so thick it was like driving through a wall, and then the next day they're paving that very road, and really paving it, with two or three layers. So: that's where they're waiting for the big bosses. Later on I filmed them, walking very nice and straight on that fresh asphalt. Not a centimeter to the side. I had that on film, too, but I didn't put it in the script.

No one could understand anything, that was the scariest thing. The dosimetrists gave one set of figures, the newspapers gave another. So gradually I begin to understand something myself: I have a little kid at home, and my dear wife . . . what kind of idiot do I have to be to be here? Well, maybe they'll give me a medal. But my wife will leave me. The only salvation was in humor. There were all sorts of jokes. In one village the only people left were a bum and four women. "So, how's your husband?" they'd say to each other. "Oh, that scoundrel runs over to the other village, too." If you tried to be serious about it all the time—Chernobyl—they're paving the road—the stream is still running, just running. But this has happened. I've felt something like this when someone close to me died. The sun is out, and the birds are flying, and the swallows, it starts raining—but he's dead. Do you understand? I want to explain this whole other dimension in a few words, explain how it was for me then.

I started filming the apple trees in bloom. The bumble bees are buzzing, there's a white, bridal color. Again, people are working, the gardens are in bloom. I'm holding the camera in my hands, but I don't understand it. This isn't right! The exposure is normal, the picture is pretty, but something's not right. And then it hits me: I don't smell anything. The garden is blooming, but there's no smell! I learned later on that sometimes the body reacts to high doses of radiation by blocking the function of certain organs. At the time I thought of my mother, who's seventy-four and can't smell, and I figured this had happened to me too. I asked the others, there were three of us: "How do the apple trees smell?" "They don't smell like anything." Something was happening to us. The lilacs didn't smell—lilacs! And I got this sense that everything around me was fake. That I was on a film set. And that I couldn't understand it. I'd never even read about anything like it.

When I was a kid, the neighbor woman, she'd been a partisan during the war, she told me a story about how their unit was surrounded but they escaped. She had her little baby with her, he was one month old, they were moving along a swamp, and there were Germans everywhere. The baby was crying. He might have given them away, they would have been discovered, the entire unit. And she suffocated him. She talked about this distantly, as if it hadn't been her, and the child wasn't hers. I can't remember now why she told me this. What I remember very clearly is my horror. What had she done? How could she? I thought the whole unit was getting out from the encirclement for that little baby, to save him. Whereas here, in order to save the life of strong healthy men, they choked this child. Then what's the point of life? I didn't want to live after that. I was a boy but I felt uncomfortable looking at this woman after I'd found this out about her.

And how did she see me? [*Silent for a while.*] Here's why I don't want to remember those days I spent in the Zone. I invent various explanations, but I don't want to open that door. I wanted to understand there what about me was real and what about me was unreal.

One night in my hotel, I wake up to this monotonous sound out my window, and strange blue lights. I pull open the curtains: dozens of trucks with red crosses and sirens are moving down the street in complete silence. I experienced something like shock. I remembered snippets from a film from my childhood. Growing up after the war we loved the war films, and there these parts, this feeling, where if everyone's left town, and you're the only good one left, what do you do? What's the right thing? Do you pretend you're dead? Or what?

In Khoyniki, there was a "plaque of achievement" in the center of town. The best people in the region had their names on it. But it was the alcoholic cab driver who went into the radioactive zone to pick up the kids from kindergarten, not any of the people on the plaque. Everyone became what he really was. And that's another thing: the evacuation. They moved the children out first, loaded them onto these big buses. And suddenly I catch myself filming everything just the way I saw it filmed in the war movies. And then I notice that the people are behaving the same way. They're all carrying themselves just like in that scene from everyone's favorite movie, *The Cranes Are Flying*—a lone tear, short words of farewell. It turned out we were all looking for a form of behavior that was familiar to us. We wanted to live up to the moment, and this is what we remembered. The girl is waving to her mom in a way that says, "Everything's fine, I'm brave. We'll win!"

I thought that I'd get to Minsk and they'd be evacuating there, too. How will I say goodbye to my wife and son? And I imagined myself making that same gesture: we'll win! We're warriors. As far back as I can remember, my father wore military clothing, though he wasn't in the military. Thinking about money was bourgeois, thinking about your own life was unpatriotic. The normal state of life was hunger. They, our parents, lived through a great catastrophe, and we needed to live through it, too. Otherwise we'd never become real people.

That's how we're made. If we just work each day and eat well—that would be strange and intolerable!

We lived in the dormitory of some technical institute with the liquidators. They were young guys. They gave us a suitcase of vodka. It helps get rid of the radiation. Suddenly we learn that there's a crew of nurses in the dormitory. All girls. "Ah, now we'll have some fun," the guys say. Two of them go over and come back with eyes like this popping out of their heads. These girls are walking along the hallways. Under their pajamas they have pants and long johns with strings, they drag on the floor behind them, they're just loose, no one cares. Everything's old, used, nothing fits. It hangs on them like on hangers. Some of them are in slippers, some are in old boots that are falling apart. And on top of all this they're wearing these half-rubber outfits that have been treated by chemicals somehow, and some of them don't take these off even at night. It's a horrible sight. And they're not nurses, either, they just pulled them out of the institute, from the military studies department. They were told it was for the weekend, but when we got there they'd already been in the Zone a month. They told us that they'd been taken to the reactor and had looked at burns, but they were the only ones who talked to us about burns. I can still picture them, going through the dormitory like sleepwalkers.

In the papers they wrote that luckily the wind was blowing in the other direction, not toward the city, not toward Kiev. All right. But it was blowing toward Belarus, toward me and my Yurik. We were walking through the forest that day, picking at some cabbage. God, how could no one warn me? We came back from the forest to Minsk. I'm riding to work on the bus, and I overhear snatches of conversation: they were filming in Chernobyl, and one cameraman died right there. He burned up. I'm wondering who it was and whether I know him. Then I hear: a young guy, two kids. They say his name: Vitya Gurevich. We do have a cameraman by that name, a real young guy. But two kids? Why didn't he tell us? We get closer to the studio, and someone corrects the information: it's not Gurevich, it's Gurin, Sergei. God, but that's me! It's funny now, but I walked to the studio worrying that I was going to open the door and see a memorial to me with my photo on it. And then this absurd thought: "Where'd they get my photograph? In the human resources department?"

Where did that rumor come from? I think from the incongruity of the scale of the event with the number of victims. For example, the Battle of Kursk—thousands dead, that was something you could understand. But here, in the first few days it was something like seven firemen. Later, a few more. But after that, the definitions were too abstract for us to understand: "in several generations," "forever," "nothing." So there were rumors: three-headed birds, chickens pecking foxes to death, bald hedgehogs. Well, and so on. Then they needed someone else to go to the Zone. One cameraman brings in a certificate saying he has an ulcer, another is on vacation. They call me in. "You have to go again." "But I just got back." "That's the thing, you were already there, so it doesn't matter to you. Besides, you already have kids. Whereas the other guys are still young." Ah, Christ, maybe I want to have five or six kids! But they start to pressure me, like, soon we re-evaluate the salaries, you'll have this under your belt, you'll get a raise. It's a sad and funny story. I've already pretty much put it away in a corner of my mind.

One time I filmed people who'd been in concentration camps. They try to avoid meeting one another. I understand that. There's something unnatural about getting together and remembering the war. People who've been through that kind of humiliation together, or who've seen what people can be like, at the bottom, run from one another. There's something I felt in Chernobyl, something I understood that I don't really want to talk about. About the fact, for example, that all our humanistic ideas are relative. In an extreme situation, people don't behave the way you read about in books. Sooner the other way around. People aren't heroes.

We're all—peddlers of the apocalypse. Big and small. I have these images in my mind, these pictures. The chairman of the kolkhoz wants two cars so that he can transport his family with all its clothes and furniture, and so the Party organization wants a car, too, it demands fairness. Meanwhile, I've seen that for several days they don't have enough cars to transport a group from the nursery school. And here two cars aren't enough to pack up all their things, including the three-liter cans of jam and pickled vegetables. I saw how they packed them up the next day. I didn't shoot that, either. [*Laughs suddenly.*] We bought some salami, some canned food, in the store, but we were afraid to eat it. We drove it around with us, though, because we didn't want to throw it out. [*Serious now.*] The mechanism of evil will work under conditions of apocalypse, also. That's what I understood. Man will gossip, and kiss up to the bosses, and save his television and ugly fur coat. And people will be the same until the end of time. Always.

I feel bad that I wasn't able to get our filming group any benefits afterwards. One of our guys needed an apartment, so I went to the union committee. "Help us out, we were in the Zone for six months. We should receive some benefits." "All right," they said, "bring us your certificates. You need certificates, with seals." But we'd gone to the regional committee in the Zone and there was just one lady there, Nastya, going around with a mop. Everyone had run off. There was a director here, he had a whole stack of certificates: where he was, what he'd filmed. A hero!

I have this big, long film in my memory, the one I didn't make. It's got many episodes. [*Silent.*] We're all peddlers of the apocalypse.

One time we went with the soldiers into a hut, there's an old lady living there.

"All right, grandma, let's go."

"Sure, boys."

"Then get your things together, grandma."

We wait outside, smoking. And then this old lady comes out: she's carrying an icon, a cat, and a little bundle in a knot. That's all she's bringing.

"Grandma, you can't bring the cat. It's not allowed. His fur is radioactive."

"No, boys, I won't go without the cat. How can I leave him? I won't leave him by himself. He's my family."

Well, with that old lady, and with that apple tree that had no smell, that's when I started. Now I only film animals. I once showed my Chernobyl films to children, and people were mad at me: why'd you do it? They don't need to see that. And so the children live in this fear, amid all this talk, their blood is changing, their immune systems are disrupted. I was hoping five or ten people would come; we filled the whole theater. They asked all sorts of questions, but one really cut into my memory. This boy, stammering and blushing, you could tell he was one of the quiet ones, asked: "Why couldn't anyone help the animals?" This was already a person from the future. I couldn't answer that question. Our art is all about the sufferings and loves of people, but not of everything living. Only humans. We don't descend to their level: animals, plants, that other world. And with Chernobyl man just waved his hand at everything.

I searched, I asked around, I was told that in the first months after the accident, someone came up with a project for evacuating the animals along with the people. But how? How do you resettle them? Okay, maybe you could move the ones that were above the earth, but what about the ones who were *in* the earth—the bugs and worms? And the ones in the sky? How do you evacuate a pigeon or a sparrow? What do you do with them? We don't have any way of giving them the necessary information. It's also a philosophical dilemma. A perestroika of our feelings is happening here.

I want to make a film called "Hostages," about animals. A strange thing happened to me. I became closer to animals. And trees, and birds. They're closer to me than they were, the distance between us has narrowed. I go to the Zone now, all these years, I see a wild boar jumping out of an abandoned human house, and then an elk. That's what I shoot. I want to make a film, to see everything through the eyes of an animal. "What are you shooting?" people say to me. "Look around you. There's a war on in Chechnya." But Saint Francis preached to the birds. He spoke to them as equals. What if these birds spoke to him in their bird language, and it wasn't he who condescended to them?

*Sergei Gurin, cameraman*

A SCREAM

Stop, good people! We have to live here! You talk and leave, but we have to live here!

Here, I have the medical cards right in front of me. Every day I have them. I take them into my hands—every day!

Anya Budai—born 1985—380 becquerels.

Vitya Grinkevich—born 1986—785 becs.

Nastya Shablovskaya—born 1986—570 becs.

Alyosha Plenin—born 1985—570 becs.

Andrei Kotchenko—born 1987—450 becs.

They say this is impossible? And how can they live with this in their thyroids? But has anyone ever run this sort of experiment before? I read and I see, every day. Can you help? No!

Then why did you come here? To ask questions? To touch us? I refuse to trade on their tragedy. To philosophize. Leave us alone, please. We need to live here.

*Arkady Pavlovich Bogdankevich,*
*rural medical attendant*

MONOLOGUE ABOUT A NEW NATION

*Speakers: Nina Konstantinovna and Nikolai Prokhorovich Zharkov,*
*both teachers. He teaches labor studies, she teaches literature.*

*She:*
I hear about death so often that I don't even notice anymore. Have you ever heard kids talk about death? My seventh-graders argue about it: is it scary or not? Kids used to ask: where do we come from? How are babies made? Now they're worried about what'll happen after the nuclear war. They don't like the classics anymore, I read them Pushkin from memory and all I see are cold, distant stares. There's a different world around them now. They read fantasy books, this is fun for them, people leaving the earth, possessing cosmic time, different worlds. They can't be afraid of death in the way that adults are afraid of death, but death interests them as something fantastical.

I wonder about this—when death's around it forces you to think. I teach Russian literature to kids who are not like the kids I taught ten years ago. They are constantly seeing someone or something get buried, get placed underground. Houses and trees, everything gets buried. If they stand in line for fifteen, twenty minutes, some of them start fainting, their noses bleed. You can't surprise them with anything and you can't make them happy. They're always tired and sleepy. Their faces are pale and gray. They don't play and they don't fool around. If they fight or accidentally break a window, the teachers are glad about it. We don't yell at them, because they're not like kids. And they're growing so slowly. You ask them to repeat something during a lesson, and the child can't, it gets to the point where you simply ask him to repeat a sentence, and he can't. You want to ask him, "Where are you? Where?"

I think about it a lot. It's like I'm painting with water on a wall, no one knows what I'm painting, no one can guess, no one has any idea. Our life revolves around Chernobyl. Where were you when it happened, how far from the reactor did you live? What did you see? Who died? Who left? Where did they go? I remember during the first months the restaurants started buzzing again—"you only live once," "if we're going to die, let's do it to music." The soldiers came and the officers came. But now Chernobyl is with us every day. A young pregnant woman died suddenly, without any diagnosis, the pathologist didn't give a diagnosis. A little girl hanged herself, she was in fifth grade. Just . . . for no reason. A little girl. There's one diagnosis for everything—Chernobyl. No matter what happens, everyone says: Chernobyl. People get mad at us: "You're sick because you're afraid. You're sick from fear. Radiophobia." But then why do little kids get sick and die? They don't know fear, they don't understand it yet.

I remember those days. My throat was burning, there was a heaviness in my whole body. "You're hypochondriacs," the doctor told me. "Everyone's that way now because of Chernobyl." "What hypochondria? Everything hurts, I feel weak." My husband and I were too shy to admit it to one another, but our legs were beginning to go numb. Everyone complained, our friends, everyone, that you'd be walking down the street and you'd just want to lie down right there. Students would lie down on their desks and lose consciousness in the middle of class. And everyone became unhappy, gloomy, not a single kind face all day, no one smiling, nothing. From eight in the morning to nine at night the kids had to stay in the school building, it was strictly forbidden to go outside and run around.

They were given clothes: the girls got skirts and blouses, the boys got suits. But then they went home in these clothes, and what happened after that we had no idea. According to the instructions, mothers were supposed to launder the clothes every day, so the kids could come to school in clean things. But first of all, they were only given one outfit, one skirt and one blouse, and second of all, mothers were already loaded down with housework—chickens, cows, pigs, and finally they don't understand why they should launder the things every day. Dirt for them is ink, or earth, or oil stains, not isotopes with short half-lives. When I tried to explain any of this to the parents, I don't think they understood it any better than if I'd been a shaman from an African tribe. "And what is this radiation? You can't hear it and you can't see it . . . Okay, I'll tell you about radiation: I don't have enough money paycheck to paycheck. The last three days we live on milk and potatoes. Okay?" And the mother says forget it. Because you're not supposed to drink milk. And you're not supposed to eat potatoes. The government brought some Chinese stir-fry and buckwheat into the stores, but where are these people supposed to get the money for it? We get some compensation for living here—death compensation—but it's nothing, enough for two cans of food. The laundry instructions are for a certain kind of person, for a certain kind of domestic situation. But we don't have that situation! We don't have that kind of person! And then, it's not that easy to explain the difference between becs and roentgen.

From my point of view—I think of it as fatalism, as a slight fatalism. For example, you weren't allowed to use anything from your garden in the first year, but people ate it anyway, cooked it and everything. They'd planted everything so well! Try telling people that they can't eat cucumbers and tomatoes. What do you mean, "can't"? They taste fine. You eat them, and your stomach doesn't hurt. And nothing "shines" in the dark. Our neighbors put down a new floor that year from the local forest, and then they measured it, its background radiation was a hundred times over the limit. No one took that floor apart, they just kept living there, figuring everything would turn out fine, somehow, without their help, without their participation. In the beginning people would bring some products over to the dosimetrist, to check them—they were way over the limit, and eventually people stopped checking. "See no evil, hear no evil. Who knows what those scientists will think up!" Everything went on its way: they turned over the soil, planted, harvested. The unthinkable happened, but people lived as they'd lived. And being refused cucumbers from their own garden was more important than Chernobyl. The kids were kept in school all summer, the soldiers washed it with a special powder, they took off a layer of soil around the school. And in the fall? In the fall they sent the students to gather the beet-roots. Students were brought in, tech-voc types, to work the fields. Everyone was chased off. Chernobyl isn't as bad as leaving potatoes in the field.

Who's to blame? Well, who but us?

Before, we didn't notice this world around us. It was there, like the sky, like the air, as if someone had given it to us forever, and it didn't depend on us. It'll always be there. I used to lie in the forest and stare up at the sky, I'd feel so good I'd forget my own name. And now? The forest is still pretty, there're plenty of blueberries, but no one picks them anymore. In autumn, it's very seldom you hear a human voice in the forest. The fear is in our feelings, on a subconscious level. We still have our television and our books, our imagination. Children grow up in their houses, without the forest and the river. They can only look at them. These are completely different children. And I go to them and recite Pushkin, who I thought was eternal. And then I have this terrible thought: what if our entire culture is just an old trunk with a bunch of stale manuscripts? Everything I love . . .

*He:*
You know, we all had a military upbringing. We were oriented toward blocking and liquidating a nuclear attack. We needed to be ready for chemical, biological, and atomic warfare. But not to draw radionuclides out of our organisms.

You can't compare it to a war, not exactly, but everyone compares it anyway. I lived through the Leningrad Blockade as a kid, and you can't compare them. We lived there like it was the front, we were constantly being shot at. And there was hunger, several years of hunger, when people were reduced to their animal instincts. Whereas here, why, please, go outside to your garden and everything's blooming! These are incomparable things. But I wanted to say something else—I lost track—it slipped away. A-ah. When the shooting starts, God help everyone! You might die this very second, not some day in the future, but right now. In the winter there is hunger. In Leningrad people burned furniture, everything wooden in our apartment we burned, all the books, I think, we even used some old rags for the stove. A person is walking down the street, he sits down, and the next day you walk by and he's still sitting there, that is to say he froze, and he might sit there like that another week, or he might sit until the spring. Until it warms up. No one has the strength to break him out of the ice. Sometimes if someone fell on the ice someone would come up and help him. But usually they'd walk past. Or crawl past. I remember people didn't walk, they crawled, that's how slowly they were going. You can't compare that with anything!

My mother still lived with us when the reactor blew up, and she kept saying: "We've already lived through the worst thing, son. We lived through the Blockade. There can't be anything worse than that."

We were preparing for war, for nuclear war, we built nuclear shelters. We wanted to hide from the atom as if we were hiding from shrapnel. But the atom is everywhere. In the bread, in the salt. We breathe radiation, we eat it. That you might not have bread or salt, and that you might get to the point where you'll eat anything, you'll boil a leather belt so that you can feed on the smell—that I could understand. But this I can't. Everything's poisoned? Then how can we live? In the first few months there was fear. The doctors, teachers, in short, the intelligentsia, they all dropped everything and left. They just hightailed it out of here. But military discipline—give up your Party card—they weren't letting anyone out. Who's to blame? In order to answer the question of how to live, we need to know who's to blame. Well, who? The scientists or the personnel at the station? The director? The operators on duty? Tell me, why do we not do battle with automobiles as the workings of the mind of man, but instead do battle with the reactor? We demand that all atomic stations be closed, and the nuclear scientists be put in jail? We curse them! But knowledge, knowledge by itself, can't be criminal. Scientists today are also victims of Chernobyl. I want to live after Chernobyl, not die after Chernobyl. I want to understand.

People have a different reaction now. Ten years have gone by, and people measure things in terms of the war. The war lasted four years. So it's like we've gone through two wars. I'll tell you what kind of reactions people have: "Everything's passed." "Everything will turn out all right." "Ten years have gone by. We're not scared anymore." "We're all going to die! We're all going to die soon!" "I want to leave the country." "They need to help us." "Aw, to hell with it. We need to live." I think I've covered all of them. That's what we hear every day. In my opinion—we're the raw materials for a scientific experiment, for an international laboratory. There are ten million Belarussians, and two million of us live on poisoned land. It's a huge devil's laboratory. Write down the data, experiment all you want. People come to us from everywhere, they write dissertations, from Moscow and Petersburg, from Japan and Germany and Austria. They're preparing for the future. [*There's a long pause in the conversation.*]

What was I thinking about just now? I was drawing out the comparison again. I was thinking that I can talk about Chernobyl, and I can't talk about the Blockade. They sent an invitation for a meeting of "The Children of Blockaded Leningrad," and I went, but I couldn't squeeze a single word out of myself while I was there. Just tell about the fear? That's not enough. Just about the fear—at home we never talked about the Blockade, my mother didn't want us to remember it. But we talk about Chernobyl. No. [*Stops.*] We don't talk about it with each other, it's a conversation we have when someone comes here: foreigners, journalists, relatives who don't live here. Why don't we talk about Chernobyl? In school, for example? With our students? They talk about it with them in Austria, France, Germany, when they go there for medical care. I ask the kids, what did people talk about with you, what interested them? They often don't remember the cities or villages, or the last names of the people they stayed with, but they remember the presents they got and the delicious foods. Someone got a cassette player, someone else didn't. They come back in clothes that they didn't earn and their parents didn't earn, either. It's like they've been on exhibit. They keep waiting for someone to take them there again. They'll show them off again, then give them presents. They get used to it. It's already a way of living, a way of seeing the world. After that big experience "abroad," after this expensive exhibit they have to go to school. Sit in class. I can already see that these are observers. I bring them to my studio, my wooden sculptures are there. The kids like them. I say: "You can make something like this out of a simple piece of tree. Try it yourself." Wake up! It helped me get out of the Blockade, it took me years to get out.

We're often silent. We don't yell and we don't complain. We're patient, as always. Because we don't have the words yet. We're afraid to talk about it. We don't know how. It's not an ordinary experience, and the questions it raises are not ordinary. The world has been split in two: there's us, the Chernobylites, and then there's you, the others. Have you noticed? No one here points out that they're Russian or Belarussian or Ukrainian. We all call ourselves Chernobylites. "We're from Chernobyl." "I'm a Chernobylite." As if this is a separate people. A new nation.

MONOLOGUE ABOUT WRITING CHERNOBYL

The ants are crawling along the tree branch. There's military hardware everywhere. Soldiers, cries, curses, swearing, helicopters rattling. But they're crawling.

I was coming back from the Zone and, of all the things I saw that day, the only one that remained clear in my memory was the image of those ants. We'd stopped in the forest and I stood smoking next to a birch. I stood very close, leaning on it. Right in front of my face the ants were crawling on the branch, not paying us any mind. We'll be gone, and they won't notice. And me? I'd never looked at them so closely before.

At first everyone said, "It's a catastrophe," and then everyone said, "It's nuclear war." I'd read about Hiroshima and Nagasaki, I'd seen documentary footage. It's frightening, but understandable: atomic warfare, the explosion's radius. I could even imagine it. But what happened to us didn't fit into my consciousness.

You feel how some completely unseen thing can enter and then destroy the whole world, can crawl in and enter you. I remember a conversation with this scientist: "This is for thousands of years," he explained. "The decomposition of uranium: that's 238 half-lives. Translated into time: that's a billion years. And for thorium: it's fourteen billion years." Fifty, one hundred, two hundred. But beyond that? Beyond that my consciousness couldn't go. I couldn't even understand anymore: what is time? Where am I?

To write about that now, when only ten years have gone by. Write about it? I think it's senseless. You can't explain it, you can't understand it. We'll still try to imagine something that looks like our own lives now. I've tried it and it doesn't work. The Chernobyl explosion gave us the mythology of Chernobyl. The papers and magazines compete to see who can write the most frightening article. People who weren't there love to be frightened. Everyone read about mushrooms the size of human heads, but no one actually found them. So instead of writing, you should record. Document. Show me a fantasy novel about Chernobyl—there isn't one! Because reality is more fantastic.

I keep a separate notebook. I write down conversations, rumors, anecdotes. It's the most interesting thing, and it's outside of time. What remains of ancient Greece? The myths of ancient Greece.

Here's my notebook.

"For three months now the radio has been saying: the situation is stabilizing, the situation is stabilizing, the situation is stab . . ."

"Stalin's old vocabulary has sprung up again: 'agents of the Western secret services,' 'the cursed enemies of socialism,' 'an undermining of the indestructible union of the Soviet peoples.' Everyone talks about the spies and provocateurs sent here, and no one talks about iodine protection. Any unofficial information is considered foreign ideology."

"Yesterday the editor cut from my article the story about a mother of one of the firemen who went to the station the night of the nuclear fire. He died of acute radiation poisoning. After burying their son in Moscow, the parents returned to their village, which was soon evacuated. In the fall they secretly made their way through the forest back to their garden and collected a bag of tomatoes and cucumbers. The mother is satisfied: 'we filled twenty cans.' Faith in the land, in their ancient peasant experience—even the death of their son can't overturn the order of things."

" 'You listen to Radio Free Europe?' my editor asks me. I don't say anything. 'I don't need alarmists on this paper. Write me up something about heroes.' "

"But hasn't the old notion of the enemy been destroyed? The enemy is invisible, and he's everywhere. This is evil in a new guise."

"Some instructors came from the Central Committee. Their route: hotel to regional Party headquarters in a car, and back, also in a car. They study the situation by reading the headlines of the local papers. They bring whole cases of sandwiches from Minsk. They boil their tea from mineral water. They brought that, too. The woman on duty at the hotel told me. People don't believe the papers, television, or radio—they look for information in the behavior of the bosses, that's more reliable."

"The most popular fable in the Zone is that Stolichnaya Vodka is the best protection against strontium and cesium."

"What should I do with my kid? I want to put him under my arm and get the hell out. But I have a Party card in my pocket. I can't do it."

"The village stores have suddenly filled up with deficit items. I heard the secretary of the regional Party give his speech: 'We'll create paradise for you on earth. Just stay and keep working. You'll be up to your neck in salami and buckwheat. You'll have everything they have in the top specialty stores.' That is, in the regional Party's buffet. Their attitude toward the people is: vodka and salami is enough for them. But, damn it all! I've never seen so many kinds of salami in a village store. I bought some imported panty hose for my wife."

"There was a month when you could buy dosimeters, and then they disappeared. You can't write about it. You also can't write about how much radioactive fallout there is. Nor can you write about the fact that only men are left in the villages, the women and children have been evacuated. All summer the men did the laundry, milked the cows, worked on the plots. And drank, of course. And fought. A world without women . . . They crossed that out. 'Don't forget, we have enemies. We have many enemies across the ocean,' the editor told me again, threateningly. And that's why we only have good things, nothing bad. But somewhere special food is prepared, and someone saw the bosses with their suitcases . . ."

"An old lady stopped me near a police block-post: 'Will you look in on my hut? It's time to dig up the potatoes, but the soldiers won't let me through.' They were transferred. A person in a vacuum, a person with nothing. They sneak into their villages through a military blockade. Through snowy forests, through swamps, at night. They get chased, caught, by helicopters, cars. 'It's like when the Germans were here,' the old-timers say."

"Saw my first looter. He was a young guy wearing two fur coats. He was proving to a military patrol that this is how he's curing his radiation sickness. When they broke him down, he finally admitted: 'The first time, it's a little scary, but after that you get used to it. Just take a shot, and off you go.' You can't let self-preservation get in the way of your instinct. Under normal circumstances that's impossible. But that's how our kind of person gets impressive things done. Including crimes."

"I went back to the village after a year. The dogs have gone wild. I found our Rex, called him, he won't come. Did he not recognize me? Or does he not want to? He's angry at us."

"During the first weeks and months everyone went quiet. There was silence. Prostration. You need to leave, but until the last day, you think, No. Your mind is incapable of understanding what's happening. I don't remember any serious conversations, but I do remember jokes. 'Now all the stores have radio-products.' 'Impotents are divided into the radioactive and the radiopassive.' And then suddenly the jokes disappeared."

*Overheard in the hospital:*
"This boy died. Yesterday he gave me some candy."

*In line at the market:*
"Oh, good people, there are so many mushrooms this year."
"They're poisoned."

"Oh, strange person. No one's forcing you to eat them. Buy them, dry them up, and take them to the market in Minsk. You'll become a millionaire."

"They picked out spots for the churches literally from heaven. The church fathers had visions. Secret rites were performed before they built the churches. But they built the nuclear power plant like a factory. Like a pigsty. They poured asphalt on for the roof. And it was melting."

"Did you read this? They caught a soldier who'd gone AWOL right near Chernobyl. He'd dug a hole for himself and lived next to the reactor. He'd eat by going to the abandoned houses—some places he'd find lard, other places some canned pickles. He laid traps for animals. He went AWOL because the older soldiers were beating the younger ones 'to death.' He saved himself—at Chernobyl."

"The first wolf-dogs have appeared, the offspring of wolves and dogs who ran away to the forest. They're bigger than wolves, they don't pay attention to flags, aren't afraid of light or people, don't respond to hunters' calls. And the feral cats have already formed packs and attack humans. They want to have revenge on us. Their memory of how they were beneath man and served him has disappeared. And for us is disappearing the boundary between what's real and what isn't."

"Some day they'll find the remains of some very strange burials. Graveyards for animals are called bio-cemeteries by scientists. These are modern-day temples. There lie thousands of dogs, cats, horses, that were shot. And not a single name."

"Yesterday my father turned eighty. The whole family gathered around the table. I looked at him and thought about how much his life had seen: the Gulag, Auschwitz, Chernobyl. One generation saw it all. But he loves to fish. When he was younger, my mother used to get mad, she'd say, 'He hasn't missed a single skirt in the entire administrative region.' And now I notice how he lowers his gaze when there's a young, pretty woman walking toward us."

"The Zone is a separate world. A different world in the midst of the rest of the world. It was invented by the Strugatsky Brothers, but literature stepped back in the face of reality."

*From rumors:*

There are camps behind Chernobyl where they're going to place those who received heavy doses of radiation. They'll keep them there a while, observe them, then bury them.

They're taking the dead out of the nearby villages in buses and straight to the graveyards, burying thousands in mass graves. Like during the Leningrad Blockade.

Several people supposedly saw a strange light in the sky above the station on the night before the explosion. Someone even photographed it. On the film it turned out to be the steam from an extra-terrestrial object.

In Minsk they've washed the trains and the inventories. They're going to transfer the whole population to Siberia. They're already fixing up the old barracks left over from Stalin's camps. They'll start with the women and children. The Ukrainians are already being shipped.

It wasn't an accident, it was an earthquake. Something happened to the earth's core. A geological explosion. Geophysical and cosmophysical forces were at work. The military knew about it beforehand, they could have warned people, but it's all very strictly kept secret there.

There are now pike in the lakes and rivers without heads or tails. Just the body floating around.

Something similar is going to start happening soon to humans. The Belarussians will turn into humanoids.

The forest animals have radiation sickness. They wander around sadly, they have sad eyes. The hunters are afraid and feel too sorry for them to shoot. And the animals have stopped being afraid of the humans. Foxes and wolves go into the villages and play with the children.

The Chernobylites are giving birth to children who have an unknown yellow fluid instead of blood. There are scientists who insist that monkeys became intelligent because they lived near radiation. Children born in three or four generations will be Einsteins. It's a cosmic experiment being carried out on us . . .

*Anatoly Shimanskiy, journalist*


## MONOLOGUE ABOUT LIES AND TRUTHS

They've written dozens of books. Fat volumes, with commentaries. But the event is still beyond any philosophical description. Someone said to me, or maybe I read it, that the problem of Chernobyl presents itself first of all as a problem of self-understanding. That seemed right. I keep waiting for someone intelligent to explain it to me. The way they enlighten me about Stalin, Lenin, Bolshevism. Or the way they keep hammering away at their "Market! Market! Free market!" But we—we who were raised in a world without Chernobyl, now live with Chernobyl.

I'm actually a professional rocketeer, I specialize in rocket fuel. I served at Baikonur [*a space launch center*]. The programs, Kosmos, Interkosmos, those took up a large part of my life. It was a miraculous time! You give people the sky, the Arctic, the whole thing! You give them space! Every person in the Soviet Union went into space with Yuri Gagarin, they tore away from the earth with him. We all did! I'm still in love with him—he was a wonderful Russian man, with that wonderful smile. Even his death seemed well-rehearsed.

It was a miraculous time! For family reasons I moved to Belarus, finished my career here. When I came, I immersed myself into this Chernobylized space, it was a corrective to my sense of things. It was impossible to imagine anything like it, even though I'd always dealt with the most advanced technologies, with outer space technologies. It's hard even to explain—it doesn't fit into the imagination—it's—[*He thinks.*] You know, a second ago I thought I'd caught it, a second ago—it makes you want to philosophize. No matter who you talk to about Chernobyl, they all want to philosophize. But I'd rather tell you about my own work. What don't we do! We're building a church—a Chernobyl church, in honor of the Icon of the Mother of God, we're dedicating it to "Punishment." We collect donations, visit the sick and dying. We write chronicles. We're creating a museum. I used to think that I, with my heart in the condition it's in, wouldn't be able to work at such a job. My first instructions were: "Here is money, divide it between thirty-five families, that is, between thirty-five widows." All the men had been liquidators. So you need to be fair. But how? One widow has a little girl who's sick, another widow has two children, and a third is sick herself, and she's renting her apartment, and yet another has four children. At night I'd wake up thinking, "How do I not cheat anyone?" I thought and calculated, calculated and thought. And I couldn't do it. We ended up just giving out the money equally, according to the list.

But my real child is the museum: the Chernobyl Museum. [*He is silent.*] Sometimes I think that we'll have a funeral parlor here, not a museum. I serve on the funeral committee. This morning I haven't even taken off my coat when a woman comes in, she's crying, not even crying but yelling: "Take his medals and his certificates! Take all the benefits! Give me my husband!" She yelled a long time. And left his medals, his certificates. Well, they'll be in the museum, on display. People can look at them. But her cry, no one heard her cry but me, and when I put these certificates on display I'll remember it.

Colonel Yaroshuk is dying now. He's a chemist-dosimetrist. He was healthy as a bull, now he's lying paralyzed. His wife turns him over like a pillow. She feeds him from a spoon. He has stones in his kidneys, they need to be shattered, but we don't have the money to pay for that kind of operation. We're paupers, we survive on what people give us. And the government behaves like a money lender, it's forgotten these people. When he dies, they'll name a street after him, or a school, or a military unit, but that's only after he dies. Colonel Yaroshuk. He walked through the Zone and marked the points of maximum radiation—they exploited him in the fullest sense of the term, like he was a robot. And he understood this, but he went, he walked from the reactor itself and then out through all the sectors around the radius of radioactivity. On foot. With a dosimeter in his hand. He'd feel a "spot" and then walk around its borders, so he could put it on his map accurately.

And what about the soldiers who worked on the roof of the reactor? Two hundred and ten military units were thrown at the liquidation of the fallout of the catastrophe, which equals about 340,000 military personnel. The ones cleaning the roof got it the worst. They had lead vests, but the radiation was coming from below, and they weren't protected there. They were wearing ordinary cheap imitation-leather boots. They spent about a minute and a half, two minutes on the roof each day, and then they were discharged, given a certificate and an award—one hundred rubles. And then they disappeared to the vast peripheries of our motherland. On the roof they gathered fuel and graphite from the reactor, shards of concrete and metal. It took about twenty to thirty seconds to fill a wheelbarrow, and then another thirty seconds to throw the "garbage" off the roof. These special wheelbarrows weighed forty kilos just by themselves. So you can picture it: a lead vest, masks, the wheelbarrows, and insane speed.

In the museum in Kiev they have a mold of graphite the size of a soldier's cap, they say that if it were real, it would weigh 16 kilos, that's how dense and heavy graphite is. The radio-controlled machines they used often failed to carry out commands or did the opposite of what they were supposed to do, because their electronics were disrupted by the high radiation. The most reliable "robots" were the soldiers. They were christened the "green robots" (by the color of their uniforms). Three thousand six hundred soldiers worked on the roof of the ruined reactor. They slept on the ground, they all tell of how in the beginning they were throwing straw on the ground in the tents—and the straw was coming from stacks near the reactor.

They were young guys. They're dying now too, but they understand that if it wasn't for them . . . These are people who came from a certain culture, the culture of the great achievement. They were a sacrifice. There was a moment when there existed the danger of a nuclear explosion, and they had to get the water out from under the reactor, so that a mixture of uranium and graphite wouldn't get into it—with the water they would have formed a critical mass. The explosion would have been between three and five megatons. This would have meant that not only Kiev and Minsk, but a large part of Europe would have been uninhabitable. Can you imagine it? A European catastrophe. So here was the task: who would dive in there and open the bolt on the safety valve? They promised them a car, an apartment, a dacha, aid for their families until the end of time. They searched for volunteers. And they found them! The boys dove, many times, and they opened that bolt, and the unit was given 7000 rubles. They forgot about the cars and apartments they promised—but that's not why they dove! Not for the material, least of all for the material promises. [*Becomes upset.*] Those people don't exist anymore, just the documents in our museum, with their names. But what if they hadn't done it? In terms of our readiness for self-sacrifice, we have no equals.

I met this one man, he was saying that this is because we place a low value on human life. That it's an Asiatic fatalism. A person who sacrifices himself doesn't feel himself to be a unique individual. He experiences a longing for his role in life. Earlier he was a person without a text, a statistic. He had no theme, he served as the background. And now suddenly he's the main protagonist. It's a longing for meaning. What does our propaganda consist of? Our ideology? You're offered a chance to die so that you can gain meaning, and be raised up. They'll give you a role! That's the high value of death, because death is eternal. This is what he proved to me, this fellow I was arguing with.

But I reject this! Categorically! Yes, we are raised to be soldiers. That's how we were taught. We're always being mobilized, always ready to do the impossible. When I finished school and wanted to go to a civilian university, my father was shocked: "I'm a career military officer, and you're going to go around in a suit jacket? The motherland needs to be protected!" He wouldn't talk to me for several months, until I put in an application to a military college. My father fought in the war; he's dead now. But he basically had no material belongings, just like the rest of his generation. He left nothing after him: no house, car, land. And what do I have of his? A field officer's bag, he got it before the Finnish campaign, his battlefield medals are in it. Also I have a bag filled with 300 letters he wrote from the front, starting in 1941, my mother saved them. That's all that's left after my father. But I consider these articles invaluable.

Now do you understand how I see our museum? In that urn there is some land from Chernobyl. A handful. And there's a miner's helmet. Also from there. Some farmer's equipment from the Zone. We can't let the dosimeters in here—we're glowing! But everything here needs to be real. No plaster casts. People need to believe us. And they'll only believe the real thing, because there are too many lies around Chernobyl. There were and there are still. They've even grown funds and commercial structures . . .

Since you're writing this book, you need to have a look at some unique video footage. We're gathering it little by little. It's not a chronicle of Chernobyl, no, they wouldn't let anyone film that, it was forbidden. If anyone did manage to record any of it, the authorities immediately took the film and returned it ruined. We don't have a chronicle of how they evacuated people, how they moved out the livestock. They didn't allow anyone to film the tragedy, only the heroics. There are some Chernobyl photo albums now, but how many video and photo cameras were broken! People were dragged through the bureaucracy. It required a lot of courage to tell the truth about Chernobyl. It still does. Believe me! But you need to see this footage: the blackened faces of the firemen, like graphite. And their eyes? These are the eyes of people who already know that they're leaving us. There's one fragment showing the legs of a woman who the morning after the catastrophe went to work on her plot of land next to the atomic station. She's walking on grass covered with dew. Her legs remind you of a grate, everything's filled with holes up to the knees. You need to see this if you're writing this book.

I come home and I can't take my little boy in my arms. I need to drink 50 or 100 grams of vodka before I can pick him up.

There's an entire section of the museum devoted to the helicopter pilots. There's Colonel Vodolazhsky, a Hero of Russia, buried on Belarussian ground in the village of Zhukov Lug. After he received more than the allowable dose of radiation, he was supposed to leave right away, but he stayed and trained thirty-three more helicopter crews. He himself performed 120 flights, releasing 230 tons of cargo. He made an average of between four and five flights per day, flying at 300 meters above the reactor, with the temperature in his cabin up to 60 degrees Celsius. Imagine what was happening below as the bags of sand were being dropped from above. The activity reached 1800 roentgen per hour; pilots began to feel it while still in the air. In order to hit the target, which was a fiery crater, they stuck their heads out of their cabins and measured it with the naked eye. There was no other way. At the meetings of government commissions, every day it was very simply said: "We'll need to put down two to three lives for this. And for this, one life." Simply, and every day.

Colonel Vodolazhsky died. On the card indicating the amount of radiation he received above the reactor, the doctors put down 7 becs. In fact it was 600!

And the four hundred miners who worked round the clock to blast a tunnel under the reactor? They needed a tunnel into which to pour liquid nitrogen and freeze the earthen pillow, as the engineers call it. Otherwise the reactor would have gone into the groundwater. So there were miners from Moscow, Kiev, Dniepropetrovsk. I didn't read about them anywhere. But they were down there naked, with temperatures reaching fifty degrees Celsius, rolling little cars before them while crouching down on all fours. There were hundreds of roentgen. Now they're dying. But if they hadn't done this? I consider them heroes, not victims, of a war, which supposedly never happened. They call it an accident, a catastrophe. But it was a war. The Chernobyl monuments look like war monuments.

There are things that aren't discussed by us, this is our Slavic modesty coming through. But you should know, since you're writing this book. Those who worked on the reactor or near it, their—they—it's a common symptom for rocketeers also, this is well-known—their urino-genital system ceases to function. But no one talks of this out loud. It's not accepted. I once accompanied an English journalist, he had put together some very interesting questions. Specifically on this theme—he was interested in the human aspect of the story—how it is for people at home, in their family life, in their intimate life. But he wasn't able to have a single honest conversation. He asked me to get together some helicopter pilots, to talk with them in the company of men. They came, some of them were already retired at the age of thirty-five, forty, one of them came with a broken leg, his bones had softened because of the radiation. But the other guys brought him. The Englishman asks them questions: how is it now with your families, with your young wives? The helicopter pilots are silent, they came to tell about their five flights a day, and he's asking about their wives? About *that*? So he starts asking them one by one, and they all answer the same: we're healthy, the government values us, and in our families all is love. Not one, not a single one of them opened up to him. They left, and I feel that he's just crushed. "Now you understand," he says to me, "why no one believes you? You lie to yourselves." The meeting had taken place in a café, and we were being served by two pretty waitresses, and he says to them: "Could you answer a few questions for me?" And they explain everything. He says, "Do you want to get married?" "Yes, but not here. We all dream of marrying a foreigner, so we can have healthy kids." And he gets braver: "Well, and do you have partners? How are they? Do they satisfy you? You understand, right, what I mean?" "You saw those guys," the waitresses say, laughing, "the helicopter pilots? Six feet tall. With their shiny medals. They're nice for meetings of the presidium, but not for bed." The Englishman photographed the waitresses and to me he repeated the same thing: "Now you understand why no one believes you? You lie to yourselves."

He and I went to the Zone. It's a well-known statistic that there are 800 waste burial sites around Chernobyl. He was expecting some fantastically engineered structures, but these were ordinary ditches. They're filled with "orange forest," which was cut down in an area of 150 hectares around the reactors. [*In the days after the accident, the pines and evergreens around the reactor turned red, then orange.*] They're filled with thousands of tons of metal and steel, small pipes, special clothing, concrete constructions. He showed me a photo from an English magazine that had a panoramic view from above. You could see thousands of individual pieces of automotive and aviation machinery, fire trucks and ambulances. The biggest graveyard is next to the reactor. He wanted to photograph it, even now, ten years later. They'd promised him more money if he got a photograph of it. So we're going around and around, from one boss to the next, one doesn't have a map, the other doesn't have permission. We ran and ran, until suddenly I realized: that graveyard no longer exists. It's just there in their account books, but it was taken apart long ago and carried off to the market, for spare parts for the kolkhoz and people's homes. Everything's been stolen and moved out. The Englishman couldn't understand this. I told him the whole truth and he didn't believe me. And even I, even when I read the bravest article, I don't believe it, I sometimes think to myself: "What if that's also a lie?" It's become a cliché to mark the tragedy. A way of greeting! A scarecrow! [*He is in despair, then is silent.*]

I drag everything to the museum. I bring it in. Sometimes I think, "Forget it! Run away!" I mean, how am I supposed to take this?

I had a conversation once with a young priest. We were standing at the grave of Sergeant-Major Sasha Goncharov. He'd worked on the roof of the reactor. It's snowing and the wind is blowing. Terrible weather. The minister is reading the mourning prayer without a hat on his head. "It's like you didn't feel the weather," I said to him afterward. "It's true," he said. "In moments like that I feel all-powerful. No church rite gives me so much energy as the mourning prayer." I remember that—the words of a man who was always near death. I've often asked foreign journalists, some of whom have been here many times, why they come, why they ask to get into the Zone? It would be silly to think it was just for money or for their careers. "We like it here," they say, "we get a real burst of life-energy here." It's an unexpected answer, no? For them, I think, the sort of person we have here, his feelings, his world, are something undiscovered and hypnotic. But I didn't ask them to clarify whether they like us ourselves, or what they can write about us, what they can understand through us.

Why do we keep hovering around death?

Chernobyl—we won't have another world now. At first, it tore the ground from under our feet, and it flung pain at us for real, but now we realize that there won't be another world, and there's nowhere to turn to. The sense of having settled, tragically, on this land—it's a completely different worldview. People returning from the war were called a "lost" generation. We're also lost. The only thing that hasn't changed is human suffering. It's our only capital. It's invaluable!

I come home after everything—my wife listens to me—and then she says quietly: "I love you, but I won't let you have my son. I won't let anyone have him. Not Chernobyl, not Chechnya. Not anyone!" The fear has already settled into her.

*Sergei Vasilyevich Sobolev, Deputy Head of the Executive Committee of the Shield of Chernobyl Association*

PEOPLE'S CHORUS

*Klavdia Grigorievna Barsuk, wife of a liquidator; Tamara Vasilyevna Belookaya, doctor; Yekaterina Fedorovna Bobrova, transferred resident from the town of Pripyat; Andrei Burtys, journalist; Ivan Naumovich Vergeychik, pediatrician; Yelena Ilyinichna Voronko, resident of the settlement of Bragin; Svetlana Govor, wife of a liquidator; Natalya Maksimovna Goncharenko, transferred resident; Tamara Ilyinichna Dubikovskaya, resident of the settlement of Narovlya; Albert Nikolaevich Zaritskiy, doctor; Aleksandra Ivanovna Kravtsova, doctor; Eleonora Ivanovna Ladutenko, radiologist; Irina Yurevna Lukashevich, midwife; Antonina Maksimovna Larivonchik, transferred resident; Anatoly Ivanovich Polischuk, hydro-meteorologist; Maria Yakovlevna Saveleyeva, mother; Nina Khantsevich, wife of a liquidator.*

It's been a long time since I've seen a happy pregnant woman. A happy mother. One gave birth recently, as soon as she got herself together she called, "Doctor, show me the baby! Bring him here." She touches the head, forehead, the little body, the legs, the arms. She wants to make sure: "Doctor, did I give birth to a normal baby? Is everything all right?" They bring him in for feeding. She's afraid: "I live not far from Chernobyl. I went there to visit my mother. I got caught under that black rain."

She tells us her dreams: that she's given birth to a calf with eight legs, or a puppy with the head of a hedgehog. Such strange dreams. Women didn't used to have such dreams. Or I never heard them. And I've been a midwife for thirty years.

*

I'm a schoolteacher, I teach Russian. This happened, I think, in early June, during exams. The director of the school suddenly gathers us all together and announces, "Tomorrow, everyone bring your shovels with you." It turns out we're supposed to take off the top, contaminated layer of soil from around the school, and later soldiers will come and pave it. The teachers have questions: "What sort of protective gear will they provide us? Will they bring special outfits, respirators?" The answer is no. "Take your shovels and dig." Only two young teachers refused, the rest went out and shoveled. A feeling of oppression but also of carrying out a necessary task—that lives within us, the need to be where it's difficult and dangerous, to defend the motherland. Did I teach my students anything but that? To go, throw yourself on the fire, defend, sacrifice. The literature I taught wasn't about life, it was about war: Sholokhov, Serafimovich, Furmanov, Fadeev, Boris Polevoy. Only two young teachers refused. But they're from the new generation. These are already different people.

We were out there digging from morning to night. When we came home, it was strange to find that the stores were open, women were buying panty hose and perfume. We already felt like it was wartime. It made a lot more sense when there suddenly appeared lines for bread, salt, matches. Everyone rushed to dry their bread into crackers. This seemed familiar to me, even though I was born after the war. I could imagine how I'd leave my house, how the kids and I would leave, which things we'd take with us, how I'd write my mother. Although all around life was going on as before, the television was showing comedies. But we always lived in terror, we know how to live in terror, it's our natural habitat. In this our people have no peers.

*

The soldiers would enter a village and evacuate the people. The village streets filled up with military hardware: APCs, large trucks with green canvas tarps, even tanks. People left their homes in the presence of soldiers, which is an oppressive atmosphere, especially for those who'd been through the war. At first they blamed the Russians—it's their fault, it was their station. Then: "The Communists are to blame."

It was constantly being compared to the war. But this was bigger. War you can understand. But this? People fell silent.

*

It was as if I'd never gone anywhere. I walk through my memories each day. Along the same streets, past the same houses. It was such a quiet town.

It was a Sunday, I was lying out, getting a tan. My mother came running: "My child, Chernobyl blew up, people are hiding in their homes, and you're lying here in the sun!" I laughed: it's forty kilometers from Chernobyl to Narovlya.

That evening a Zhiguli stops in front of our house and my friend and her husband come in. She's wearing a bathrobe and he's in an athletic suit and some old slippers. They went through the forest, along some tiny village roads, from Pripyat. The roads were being patrolled by police, military block-posts, they weren't letting anyone out. The first thing she yelled was: "We need to find milk and vodka! Hurry!" She was yelling and yelling. "We just bought new furniture, a new refrigerator. I sewed myself a fur. I left everything, I wrapped it in cellophane. We didn't sleep all night. What's going to happen? What's going to happen?" Her husband tried to calm her down. We sat in front of the television for days, waiting for Gorbachev to speak. The authorities didn't say anything. Only after the big holiday did Gorbachev come on and say: don't worry, comrades, the situation is under control. It's nothing bad. People are still there, living, working.

*

They herded all the livestock from the evacuated villages into designated points in our regional center. These cows, calves, pigs, they were going crazy, they would run around the streets—whoever wanted to catch them could catch them. The cars with the canned meat went from the meat combine to the station at Kalinovich, and from there to Moscow. Moscow wouldn't accept the cargo. So these train cars, which were by now graveyards, came back to us. Whole echelons of them, and we buried them here. The smell of rotten meat followed me around at night. "Can it be that this is what an atomic war smells like?" The war I remembered smelled of smoke.

At first, they bused the children out at night. They were trying to hide the catastrophe. But people found out anyway. They'd bring milk cans out to our buses, they baked pies. It was just like during the war. There's nothing else to compare it to.

There was a meeting at the regional executive's office. It felt like a military situation. Everyone was waiting for the head of the civil defense to speak, because no one remembered anything about radiation aside from some passages from their tenth-grade physics textbook. He goes out on stage and begins to tell us what's written in the books about nuclear warfare: that once a soldier has taken 50 roentgen, he must leave the field; how to build a shelter; how to put on a gas mask; facts about the radius of the explosion.

We went into the contaminated zone on a helicopter. We were all properly equipped—no undergarments, a raincoat out of cheap cotton, like a cook's, and that's covered with a defensive field, then mittens, and a gauze surgical mask. We have all sorts of instruments hanging off us. We come out of the sky near a village and we see that there are boys playing in the sand, like nothing had happened. One has a rock in his mouth, another a tree branch. They're not wearing pants, they're naked. But we have orders, not to panic the population.

And now I live with this.

*

They suddenly started having these segments on television, like: an old lady milks her cow, pours the milk into a can, the reporter comes over with a military dosimeter, measures it. And the commentator says, See, everything's fine, and the reactor is just ten kilometers away. They show the Pripyat

River, there are people swimming in it, tanning themselves. In the distance you see the reactor and plumes of smoke above it. The commentator says: The West is trying to spread panic, telling lies about the accident. And then they show the dosimeter again, measuring some fish on a plate, or a chocolate bar, or some pancakes at an open pancake stand. It was all a lie. The military dosimeters then in use by our armed forces were designed to measure the radioactive background, not individual products.

This level of lying, this incredible level, with which Chernobyl is connected in our minds, was comparable only to the level of lies during the big war.

\*

We were expecting our first child. My husband wanted a boy and I wanted a girl. The doctors tried to convince me: "You need to get an abortion. Your husband was at Chernobyl." He was a truck driver; they called him in during the first days. He drove sand. But I didn't believe anyone.

The baby was born dead. She was missing two fingers. A girl. I cried. "She should at least have fingers," I thought. "She's a girl."

\*

No one could understand what had happened. I called military headquarters—all medical personnel have military obligations—and volunteered to help. I can't remember his name, but he was a major, and he told me, "We need young people." I tried to convince him: "Young doctors aren't ready, first of all, and second of all, they will be in greater danger because young people are more susceptible to radiation." His answer: "We have our orders, we're to take young people."

Patients' wounds began to heal more slowly. I remember that first radioactive rain—"black rain," people called it later. First off, you're just not ready for it, and second, we're the best, most extraordinary, most powerful country on Earth. My husband, a man with a university degree, an engineer, seriously tried to convince me that it was an act of terrorism. An enemy diversion. A lot of people at the time thought that. But I remembered how I'd once been on a train with a man who worked in construction who told me about the building of the Smolensk nuclear plant: how much cement, boards, nails, and sand was stolen from the construction site and sold to neighboring villages. In exchange for money, for a bottle of vodka.

People from the Party would come to the villages and the factories to speak with the populace, but not one of them could say what deactivation was, how to protect children, what the coefficient was for the leakage of radionuclides into the food supply. They didn't know anything about alpha- or beta- or gamma-rays, about radiobiology, ionizing radiation, not to mention about isotopes. For them, these were things from another world. They gave talks about the heroism of the Soviet people, told stories about military bravery, about the machinations of Western spy agencies. When I even mentioned this briefly in a Party meeting, when I doubted this, I was told that they'd kick me out of the Party.

*

I'm afraid of staying on this land. They gave me a dosimeter, but what am I supposed to do with it? I do my laundry, it's nice and white, but the dosimeter goes off. I make some food, bake a pie—it goes off. I make the bed—it goes off. What do I need it for? I feed my kids and cry. "Why are you crying, Mom?"

I have two boys. They don't go to nursery school or kindergarten—they're always in the hospital. The older one—he's neither a boy nor a girl. He's bald. I take him to the doctors, and also to the healers. He's the littlest one in his grade. He can't run, he can't play, if someone hits him by accident and he starts bleeding, he might die. He has a blood disease, I can't even pronounce the word for it. I'm lying with him in the hospital and thinking, "He's going to die." I understood later on that you can't think that way. I cried in the bathroom. None of the mothers cry in the hospital rooms. They cry in the toilets, the baths. I come back cheerful: "Your cheeks are red. You're getting better."

"Mom, take me out of the hospital. I'm going to die here. Everyone here dies."

Now where am I going to cry? In the bathroom? There's a line for the bathroom—everyone like me is in that line.

*

On May 1, on the day of memory, they let us into the cemetery. They let us go to the graves, but the police forbid us from going to our houses and our gardens. From the cemetery at least we looked at our homes from afar. We blessed them from where we were.

*

Let me tell you about the sort of people who live here. I'll give you one example. In the "dirty" areas, during the first few years, they were filling the stores with Chinese beef, and buckwheat, and everything, and people said, "Oh, it's good here. You won't get us to leave here now." The land became contaminated unevenly—one kolkhoz might have "clean" fields next to "dirty" ones. People who work on the "dirty" fields get paid more, and everyone's rearing to work there. And they refuse to work the "clean" fields.

Not long ago my brother visited me from the Far East. "You're all like black boxes here," he said. He meant the black boxes that record information on airplanes. We think that we're living, talking, walking, eating. Loving one another. But we're just recording information!

\*

I'm a pediatrician. It's different for children. For example, they don't think that cancer means death—that connection hasn't been made for them. And they know everything about themselves: their diagnosis, the medicines they're taking, the names of the procedures. They know more than their mothers. When they die, they have these surprised looks on their faces. They lie there with these surprised faces.

\*

The doctors warned me that my husband would die. He has leukemia—cancer of the blood. He got sick after he came back from the Chernobyl Zone, two months after. He was sent there from the factory. He came home one morning after the night shift:

"I'm leaving tomorrow."

"What are you going to do there?"

"Work on the kolkhoz."

They raked the straw in the fifteen-kilometer zone, collected the beets, dug up the potatoes.

He came back. We went to visit his parents. He was spackling a wall with his father when he fell down. We called an ambulance, took him to the hospital—he'd received a fatal dose of leukocytes.

He returned with one thought in his mind: "I'm dying." He became quiet. I tried to convince him it wasn't true. I begged him. He wouldn't believe me. Then I gave him a daughter, so he'd believe me. I'd wake up in the morning, look at him: how am I going to make it by myself? You shouldn't think a lot about death. I chased the thoughts away. If I'd known he'd get sick I'd have closed all the doors, I'd have stood in the doorway. I'd have locked the doors with all the locks we had.

*

We've been going from hospital to hospital with my son for two years now. I don't want to hear anything, read anything about Chernobyl. I've seen it all.

The little girls in the hospitals play with their dolls. They close their eyes and the dolls die.

"Why do the dolls die?"

"Because they're our children, and our children won't live. They'll be born and then die."

My Artyom is seven, but he looks five. He'll close his eyes, and I think he's gone to sleep. I'll start crying, since he can't see me. But then he says, "Mom, am I dying already?"

He'll go to sleep, and he's almost not breathing. I'll get on my knees before him, before his bed, "Artyom, open your eyes. Say something." And I'll think to myself, "You're still warm."

He opens his eyes, then goes back to sleep again, so quietly, as if he's dying.

"Artyom, open your eyes."

I won't let him die.

*

Not long ago we celebrated the New Year. We had everything, and it was all homemade: smoked goods, lard, meat, pickles. The only thing from the store was the bread. Even the vodka was ours. Of course "ours" meant that it was from Chernobyl. With cesium, and a strontium aftertaste. But where else are we going to get anything? The village stores are empty, and if something appears in them, we can't buy it on our salaries and our pensions.

Some guests came over, our neighbors, very nice people, young; one is a teacher, the other is a mechanic on the kolkhoz and his wife. We drank, had some food. And then we started singing. Spontaneously we sang all the old songs—the revolutionary songs, the war songs. "The morning sun colors the ancient Kremlin with its gentle light." And it was a nice evening. It was like before.

I wrote about it to my son. He's a student, he lives in the capital. He writes me back: "Mom, I pictured the scene to myself. It's crazy. That Chernobyl land, our house. The New Year's tree is sparkling. And the people at the table are singing revolutionary songs and military songs. As if they hadn't gone through the Gulag, and through Chernobyl." I became frightened—not for myself, but for my son. He has nowhere to come back to.

# PART THREE

# AMAZED BY SADNESS

MONOLOGUE ABOUT WHAT WE DIDN'T KNOW:
DEATH CAN BE SO BEAUTIFUL

At first, the question was, Who's to blame? But then, when we learned more, we started thinking, What should we do? How do we save ourselves? After coming to terms with the fact that this would not be for one year or for two, but for many generations, we began to look back, turning the pages.

It happened late Friday night. That morning no one suspected anything. I sent my son to school, my husband went to the barber's. I'm preparing lunch when my husband comes back. "There's some sort of fire at the nuclear plant," he says. "They're saying we are not to turn off the radio." I forgot to say that we lived in Pripyat, near the reactor. I can still see the bright-crimson glow, it was like the reactor was glowing. This wasn't any ordinary fire, it was some sort of shining. It was pretty. I'd never seen anything like it in the movies. That evening everyone spilled out onto their balconies, and those who didn't have them went to friends' houses. We were on the ninth floor, we had a great view. People brought their kids out, picked them up, said, "Look! Remember!" And these were people who worked at the reactor—engineers, workers, physics instructors. They stood in the black dust, talking, breathing, wondering at it. People came from all around on their cars and their bikes to have a look. We didn't know that death could be so beautiful. Though I wouldn't say that it had no smell—it wasn't a spring or an autumn smell, but something else, and it wasn't the smell of earth. My throat tickled, and tears came to my eyes.

I didn't sleep all night, and I heard the neighbors walking around upstairs, also not sleeping. They were carrying stuff around, banging things, maybe they were packing their belongings. I fought off my headache with Citramon tablets. In the morning I woke up and looked around and I remember feeling—this isn't something I made up later, I thought it right then—something isn't right, something has changed forever. At eight that morning there were already military people on the streets in gas masks. When we saw them on the streets, with all the military vehicles, we didn't grow frightened—on the contrary, it calmed us down. Since the army has come to our aid, everything will be fine. We didn't understand then that the peaceful atom could kill, that man is helpless before the laws of physics.

All day on the radio they were telling people to prepare for an evacuation: they'd take us away for three days, wash everything, check it over. The kids were told to take their school books. Still, my husband put our documents and our wedding photos into his briefcase. The only thing I took was a gauze kerchief in case the weather turned bad.

From the very first I felt that we were Chernobylites, that we were already a separate people. Our bus stopped overnight in a village; people slept on the floor in a school, others in a club. There was nowhere to go. One woman invited us to sleep at her house. "Come," she said, "I'll put down some linen for you. I feel bad for your boy." Her friend started dragging her away from us. "Are you crazy? They're contaminated!" When we settled in Mogilev and our son started school, he came back the very first day in tears. They put him next to a girl who said she didn't want to sit with him, he was radioactive. Our son was in the fourth grade, and he was the only one from Chernobyl in the class. The other kids were afraid of him, they called him "Shiny." His childhood had ended so early.

As we were leaving Pripyat there was an army column heading back in the other direction. There were so many military vehicles, that's when I grew frightened. But I couldn't shake the feeling that this was all happening to someone else. I was crying, looking for food, sleeping, hugging my son, calming him down, but inside, this constant sense that I was just an observer. In Kiev they gave us some money, but we couldn't buy anything: hundreds of thousands of people had been uprooted and they'd bought everything up and eaten everything. Many had heart attacks and strokes, right there at the train stations, on the buses. I was saved by my mother. She'd lived a long time and had lost everything more than once. The first time was in the 1930s, they took her cow, her horse, her house. The second time, there'd been a fire, the only thing she'd saved was me. Now she said, "We have to get through it. After all, we're alive."

I remember one thing: we're on the bus, everyone's crying. A man up front is yelling at his wife. "I can't believe you'd be so stupid! Everyone else brought their things, and all we've got are these three-liter bottles!" The wife had decided that since they were taking the bus, she might as well bring some empty pickling bottles for her mother, who was on the way. They had these big bulging sacks next to their seats, we kept tripping over them the whole way to Kiev, and that's what they came to Kiev with.

Now I sing in the church choir. I read the Bible. I go to church—it's the only place they talk about eternal life. They comfort a person. You won't hear those words anywhere else, and you so want to hear them.

I often dream that I'm riding through sunny Pripyat with my son. It's a ghost town now. But we're riding through and looking at the roses, there were many roses in Pripyat, large bushes with roses. I was young. My son was little. I loved him. And in the dream I've forgotten all the fears, as if I were just a spectator the whole time.

*Nadezhda Petrovna Vygovskaya,*
*evacuee from the town of Pripyat*

MONOLOGUE ABOUT THE SHOVEL AND THE ATOM

I tried to commit those days to memory. There were many new emotions—fear, a sense of tearing into the unknown, like I'd landed on Mars. I'm from Kursk. In 1969, they built a nuclear reactor nearby in the town of Kurchatov. We used to go there to buy food—the nuclear workers always received the best provisions. We used to go fishing in the pond there, right near the reactor. I thought of that often after Chernobyl.

So here's how it was: I received a notice, and, as I'm a disciplined person, I went to the military recruiter's office the next day. They went through my file. "You," they tell me, "have never gone on an exercise with us. And they need chemists out there. You want to go for twenty-five days to a camp near Minsk?" And I thought: Why not take a break from my family and my job for a while? I'll march around a bit in the fresh air.

At 11 A.M. on June 22, 1986, I came with a bundle and a toothbrush to the gathering spot. I was surprised by how many of us there were for a peacetime exercise. I started remembering scenes from war films—and what a day for it, June 22, the day the Germans invaded. All day they tell us to get in formation, then to break up, finally as it's getting dark we get on our buses. Someone gets on and says: "If you've brought liquor with you, drink it now. Tonight we'll get on the train, and in the morning we'll join our units. Everyone is to be fresh as a cucumber in the morning, and without excess baggage." All right, no problem, we partied all night.

In the morning we found our unit in the forest. They put us in formation again and called us up in alphabetical order. We received protective gear. They gave us one set, then another, then a third, and I thought, This is serious. They also gave us an overcoat, hat, mattress, pillow—all winter gear. But it was summer out, and they told us we'd be going home in twenty-five days. "Are you kidding?" says the captain who came with us, laughing. "Twenty-five days? You'll be in Chernobyl six months." Disbelief. Then anger. So they start convincing us: anyone working twenty kilometers away gets double pay, ten kilometers means triple pay, and if you're right at the reactor you get six times the pay. One guy starts figuring that in six months he'll be able to roll home in a new car, another wants to run off but he's in the army now. What's radiation? No one's heard of it. Whereas I've just gone through a civil defense course where they gave us information from thirty years before, like that 50 roentgen is a fatal dose. They taught us how to fall down so that the wave of the explosion would miss us. They taught us about irradiation, thermal heat. But about the radioactive contamination of an area—the most dangerous factor of all—not a word.

And the staff officers who took us to Chernobyl weren't terribly bright. They knew one thing: you should drink more vodka, it helps with the radiation. We stayed near Minsk for six days, and for all six days we drank. I studied the labels on the bottles. At first we drank vodka, and then I see we're drinking some strange stuff: Nithinol and other glass cleansers. For me, as a chemist, this was interesting. After the nithinol, your legs feel cottony but your head is clear, you give yourself a command, "Stand up!," but you fall down.

So here's how it was: I'm a chemical engineer, I have a master's degree, I was working as the head of a laboratory at a large production facility. And what did they have me do? They handed me a shovel—this was practically my only instrument. We immediately came up with a slogan: Fight the atom with a shovel! Our protective gear consisted of respirators and gas marks, but no one used them because it was 30 degrees Celsius outside, if you put those on it would kill you. We signed for them, as you would for supplementary ammunition, and then forgot all about it. It was just one more detail.

They transferred us from the buses to the train. There were forty-five seats in the train-car and seventy of us. We took turns sleeping.

So what is Chernobyl? A lot of military hardware and soldiers. Wash posts. A real military situation. They placed us in tents, ten men to a tent. Some of us had kids at home, some had pregnant wives, others were in between apartments. But nobody complained. If we had to do it, we had to do it. The motherland called and we went. That's just how we are.

There were enormous piles of empty tin cans around the tents. The military depots have a special supply in case of war. The cans were from canned meat, pearl buckwheat, sprats. There were groups of cats all around, they were like flies. The villages had been emptied—you'd hear a gate open and turn around expecting a person, and instead there'd be a cat walking out.

We dug up the diseased top layer of soil, loaded it into automobiles and took it to waste burial sites. I thought that a waste burial site was a complex, engineered construction, but it turned out to be an ordinary pit. We picked up the earth and rolled it, like big rugs. We'd pick up the whole green mass of it, with grass, flowers, roots. And bugs, and spiders, worms. It was work for madmen. You can't just pick up the whole earth, take off everything living. If we weren't drinking like crazy every night, I doubt we'd have been able to take it. Our psyches would have broken down. We created hundreds of kilometers of torn-up, fallow earth. The houses, barns, trees, highways, kindergartens, wells—they all remained there, naked. In the morning you'd wake up, you need to shave, but you're afraid to look in the mirror and see your own face. Because you're getting all sorts of thoughts. It's hard to imagine people moving back to live there again. But we changed the slate, we changed the roofs on houses. Everyone understood that this was useless work, and there were thousands of us. Every morning we'd get up and do it again. We'd meet an illiterate old man: "Ah, quit this silly work, boys. Have a seat at the table, eat with us." The wind would be blowing, the clouds floating. The reactor wasn't even shut down. We'd take off a layer of earth and come back in a week and start over again. But there was nothing left to take off—just some sand that had drifted in. The one thing we did that made sense to me was when some helicopters sprayed a special mixture that created a polymer film that kept the light-moving bottom-soil from moving. That I understood. But we kept digging, and digging . . .

The villages were evacuated, but some still had old men in them. To walk into an old peasant hut and sit down to dinner—just the ritual of it—a half hour of normal life. Although you couldn't eat anything, it wasn't allowed. But I so wanted to sit at the table, in an old peasant hut.

After we were done the only thing left were the pits. They were going to fill them with concrete plates and surround them with barbed wire, supposedly. They left the dump trucks, cargo trucks, and cranes they'd been using there, since metal absorbs radiation in its own way. I've been told that all that stuff has since disappeared, that is, been stolen. I believe it. Anything is possible here now.

One time we had an alarm: the dosimetrists discovered that our cafeteria had been put in a spot where the radiation was higher than where we went to work. We'd already been there two months by then. That's just how we are. The cafeteria was just a bunch of posts and these had boards nailed to them at chest height. We ate standing up. We washed ourselves from barrels filled with water. Our toilet was a long pit in a clear field. We had shovels in our hands, and not far off was the reactor.

After two months we began to understand things a little. People started saying: "This isn't a suicide mission. We've been here two months—that's enough. They should bring in others now." Major-General Antoshkin had a talk with us. He was very honest. "It's not advantageous for us to bring in a new shift. We've already given you three sets of clothing. And you're used to the place. To bring in new men would be expensive and complicated." With an emphasis on our being heroes. Once a week someone who was digging really well would receive a certificate of merit before all the other men. The Soviet Union's best grave digger. It was crazy.

These empty villages—just cats and chickens. You walk into a barn, it's filled with eggs. We'd fry them. Soldiers are ready for anything. We'd catch a chicken, put it on the fire, wash it down with a bottle of homemade vodka. We'd put away a three-liter bottle of that stuff every night in the tent. Someone'd be playing chess, another guy was on his guitar. A person can get used to anything. One guy would get drunk and fall down on his bed to sleep, other guys wanted to yell and fight. Two of them got drunk and went for a drive and crashed. They got them out from under the crushed metal with the jaws of life. I saved myself by writing long letters home and keeping a diary. The head of the political department noticed, he kept asking me what I was writing, where was I keeping it? He got my neighbor to spy on me, but the guy warned me. "What are you writing?" "My dissertation." He laughs. "All right, that's what I'll tell the colonel. But you should hide that stuff." They were good guys. I already said, there wasn't a single whiner in the bunch. Not a single coward. Believe me: no one will ever defeat us. Ever! The officers never left their tents. They'd walk around in slippers all day, drinking. Who cares? We did our digging. Let the officers get another star on their shoulder. Who cares? That's the sort of people we have in this country.

The dosimetrists—they were gods. All the village people would push to get near them. "Tell me, son, what's my radiation?" One enterprising soldier figured it out: he took an ordinary stick, wrapped some wiring to it, knocks on some old lady's door and starts waving his stick at the wall. "Well, son, tell me, how it is." "That's a military secret, grandma." "But you can tell me, son. I'll give you a glass of vodka." "All right." He drinks it down. "Ah, everything's all right here, grandma. Don't worry." And leaves.

In the middle of our time there they finally gave us dosimeters. These little boxes, with a crystal inside. Some of the guys started figuring, they should take them over to the burial site in the morning and let them catch radiation all day, that way they'll get released sooner. Or maybe they'll pay them more. So you had guys attaching them to their boots, there was a loop there, so that they'd be closer to the ground. It was theater of the absurd. These counters weren't even going, they needed to be set in motion by an initial dose of radiation. In other words, these were little toys they'd picked out of the warehouse from fifty years ago. It was just psychotherapy for us. At the end of our time there we all got the same thing written on our medical cards: they multiplied the average radiation by the number of days we were there. And they got that initial average from our tents, not from where we worked.

We got two hours to rest. I'd lie down under some bush, and see that the cherries are in bloom, big, juicy cherries, you wipe them down and eat them. Mulberry—it was the first time I'd seen it. When we didn't have work, they'd march us around. We watched Indian films about love, until three, four in the morning. Sometimes after that the cook would oversleep and we'd have undercooked buckwheat. They brought us newspapers—they wrote that we were heroes. Volunteers! There were photographs. If only we'd met that photographer . . .

The international units were nearby. There were Tatars from Kazan. I saw their internal court-martial. They chased a guy in front of the unit, if he stopped or went off to the side they'd start kicking him. He'd been cleaning houses and they'd found a bag full of stuff on him, he'd been stealing. The Lithuanians were nearby, too. After two months they rebelled and demanded to be sent back home.

One time we got a special order: immediately wash this one house in an empty village. Incredible! "What for?" "Tomorrow they're going to film a wedding there." So we got some hoses and doused the roof, trees, scraped off the ground. We mowed down the potato patch, the whole garden, all the grass in the yard. All around, emptiness. The next day they bring the bride and groom, and a busload of guests. They had music. And they were a real bride and groom, they weren't actors—they'd already been evacuated, they were living in another place, but someone convinced them to come back and film the wedding here, for history. Our propaganda in motion. A whole factory of daydreams. Even here our myths were at work, defending us: see, we can survive anything, even on dead earth.

Right before I went home the commander called me in. "What were you writing?" "Letters to my young wife." "All right. Be careful."

What do I remember from those days? A shadow of madness. How we dug. And dug. Somewhere in my diary I wrote down that I understood, in the first few days I understood—how easy it is to become earth.

*Ivan Nikolaevich Zhykhov, chemical engineer*

MONOLOGUE ABOUT TAKING MEASUREMENTS

Already by the end of May, about a month after the accident, we began receiving, for testing, products from the thirty-kilometer zone. The institute worked round the clock, like it was a military institute. At the time we were the only ones in Belarus with the specialists and the equipment for the job.

They brought us the insides of domestic and undomesticated animals. We checked the milk. After the first tests it became clear that what we were receiving couldn't properly be called meat—it was radioactive byproducts. Within the zone the herds were taken care of in shifts—the shepherds would come and go, the milkmaids were brought in for milking only. The milk factories carried out the government plan. We checked the milk. It wasn't milk, it was a radioactive byproduct.

For a long time after that we used dry milk powder and cans of condensed and concentrated milk from the Rogachev milk factory in our lectures as examples of a standard radiation source. And in the meantime, they were being sold in the stores. When people saw that the milk was from Rogachev and stopped buying it, there suddenly appeared cans of milk without labels. I don't think it was because they ran out of paper.

On my first trip to the Zone I measured a background radiation level of five to six times higher in the forest than on the roads or the fields. But high doses were everywhere. The tractors were running, the farmers were digging on their plots. In a few villages we measured the thyroid activity for adults and children. It was one hundred, sometimes two and three hundred times the allowable dosage. There was a woman in our group, a radiologist. She became hysterical when she saw that children were sitting in a sandbox and playing. We checked breast milk—it was radioactive. We went into the stores—as in a lot of village stores, they had the clothes and the food right next to each other: suits and dresses, and nearby salami and margarine. They're lying there in the open, they're not even covered with cellophane. We take the salami, we take an egg—we make a roentgen image—this isn't food, it's a radioactive byproduct.

We see a woman on a bench near her house, breastfeeding her child—her milk has cesium in it—she's the Chernobyl Madonna.

We asked our supervisors, What do we do? How should we be? They said: "Take your measurements. Watch television." On television Gorbachev was calming people: "We've taken immediate measures." I believed it. I'd worked as an engineer for twenty years, I was well-acquainted with the laws of physics.

I knew that everything living should leave that place, if only for a while. But we conscientiously took our measurements and watched the television. We were used to believing. I'm from the postwar generation, I grew up with this belief, this faith. Where did it come from? We'd won that terrible war. The whole world was grateful to us then.

So here's the answer to your question: why did we keep silent knowing what we knew? Why didn't we go out onto the square and yell the truth? We compiled our reports, we put together explanatory notes. But we kept quiet and carried out our orders without a murmur because of Party discipline. I was a Communist. I don't remember that any of our colleagues refused to go work in the Zone. Not because they were afraid of losing their Party membership, but because they had faith. They had faith that we lived well and fairly, that for us man was the highest thing, the measure of all things. The collapse of this faith in a lot of people eventually led to heart attacks and suicides. A bullet to the heart, as in the case of Professor Legasov, because when you lose that faith, you are no longer a participant, you're an also-ran, you have no reason to exist. That's how I understood his suicide, as a sort of sign.

*Marat Filippovich Kokhanov, former chief engineer of the Institute for Nuclear Energy of the Belarussian Academy of Sciences*


MONOLOGUE ABOUT HOW THE FRIGHTENING THINGS
IN LIFE HAPPEN QUIETLY AND NATURALLY

From the very beginning—we heard that something had happened somewhere. I didn't even hear the name of the place, but it was somewhere far from our Mogilev. Then my brother came running home from school, he said all the kids were getting pills. So apparently something really had happened.

And still we had a great time on May 1. We came home late at night, and my window had been blown open by the wind. I would remember that later on.

I worked at the inspection center for environmental defense. We were waiting for some kind of instructions, but we didn't receive any. There were very few professionals on our staff, especially among the directors: they were retired colonels, former Party workers, retirees or other undesirables. If you messed up somewhere else, they'd send you to us. Then you sit there shuffling papers. They only started making noise after our Belarussian writer Aleksei Adamovich spoke out in Moscow, raising the alarm. How they hated him! It was unreal. Their children live here, and their grandchildren, but instead of them it's a writer calling to the world: save us! You'd think some sort of self-preservation mechanism would kick in. Instead, at all the Party meetings, and during smoke breaks, all you heard about was "those writers." "Why are they sticking their noses where they don't belong? They've really let themselves go! We have instructions! We need to follow orders! What does he know? He's not a physicist! We have the Central Committee, we have the General Secretary!" I think I understood then, for the first time, a bit of what it was like in 1937. How that felt.

At that time my notions of nuclear power stations were utterly idyllic. At school and at the university we'd been taught that this was a magical factory that made "energy out of nothing," where people in white robes sat and pushed buttons. Chernobyl blew up when we weren't prepared. And also there wasn't any information. We got stacks of paper marked "Top Secret." "Reports of the accident: secret"; "Results of medical

observations: secret"; "Reports about the radioactive exposure of personnel involved in the liquidation of the accident:
secret." And so on. There were rumors: someone read in some
paper, someone heard, someone said . . . Some people listened
to what was being said in the West, they were the only ones
talking about what pills to take and how to take them. But
most often the reaction was: our enemies are celebrating, but
we still have it better. On May 9th the veterans will still go
out on their victory parade. Even those who were fighting the
fire at the reactor, as it later turned out, were living among
rumors. "I think it's dangerous to take the graphite in your
hands. I think . . ."

This crazy woman appeared in town suddenly. She'd walk
around the market saying, "I've seen the radiation. It's blue-blue,
it spills over everything." People stopped buying milk and cottage cheese at the market. An old lady would be standing with
her milk, no one's buying it. "Don't worry," she'd say, "I don't let
my cow out into the field, I bring her her grass myself." If you
drove out of town you'd see these scarecrows: a cow all wrapped
in cellophane, and then an old farmer woman next to her, also
wrapped in cellophane. You could cry, you could laugh.

And by this point they started sending us out on inspec-
tions. I was sent to a timber processing plant. They weren't
receiving any less timber—the plan hadn't been altered, so
they kept to it. I turned on my instrument at the warehouse
and it started going nuts. The boards were okay, but if I
turned it on near the brooms, it went off the chart. "Where
are the brooms from?" "Krasnopol." And Krasnopol as it
later turned out was the most contaminated place in the
Mogilev region. "We have one shipment left. The others
went out already." And how are you going to find them in
all the towns they were sent to?

There was something else I was afraid of leaving out . . . Oh, right! Chernobyl happened, and suddenly you got this new feeling, we weren't used to it, that everyone has his separate life. Until then no one needed this life. But now you had to think: what are you eating, what are you feeding your kids? What's dangerous, what isn't? Should you move to another place, or should you stay? Everyone had to make her own decisions. And we were used to living—how? As an entire village, as a collective—a factory, a kolkhoz. We were Soviet people, we were collectivized. I was a Soviet person, for example. Very Soviet. When I was in college, I went every summer with the Student Communist youth group. We'd go work for a summer, and the money was transferred to some Latin American CP. Our unit was working at least partially for Uruguay's.

Then we changed. Everything changed. It takes a lot of work to understand this. And also there's our inability to speak out.

I'm a biologist. My dissertation was on the behavior of bees. I spent two months on an uninhabited island. I had my own bee's nest there. They took me into their family after I'd spent a week hanging around. They wouldn't let anyone closer than three meters, but they were letting me walk up next to them after a week. I fed them jam off a match right into the nest. Our teacher used to say: "Don't destroy an anthill, it's a good form of alien life." A bee's nest is connected to the entire forest, and I gradually also became part of the landscape. A little mouse would come running up and sit on my sneakers—he was a wild forest mouse but he already thought I was part of the scene, I was here yesterday, I'd be there the next day.

After Chernobyl—there was an exhibit of children's drawings, one of them had a stork walking through a field, and then under it, "No one told the stork." Those were my feelings too. But I had to work. We went around the region collecting samples of water, earth, and taking them to Minsk. Our assistants were grumbling: "We're carrying hotcakes." We had no defense, no special clothing. You'd be sitting in the front seat, and behind you there were samples just glowing.

They had protocols written up for burying radioactive earth. We buried earth in earth—such a strange human activity. According to the instructions, we were supposed to conduct a geological survey before burying anything to determine that there was no ground water within four to six meters of the burial site, and that the depth of the pit wasn't very great, and also that the walls and bottom of the pit be lined with polyethylene film. That's what the instructions said. In real life it was of course different. As always. There was no geological survey. They'd point their fingers and say, "Dig here." The excavator digs. "How deep did you go?" "Who the hell knows? I stopped when I hit water." They were digging right into the water.

They're always saying: it's a holy people, and a criminal government. Well I'll tell you a bit later what I think about that, about our people, and about myself.

My longest assignment was in the Krasnopolsk region, which as I said was the worst. In order to keep the radionuclides from washing off the fields into the rivers, we needed to follow the instructions again. You had to plow double furrows, leave a gap, and then again put in double furrows, and so on with the same intervals. You had to drive along all the small rivers and check. So I get to the regional center on a bus, and then obviously I need a car. I go to the chairman of the regional executive. He's sitting in his office with his head in his hands: no one changed the plan, no one changed the harvesting operations, just as they'd planted the peas, so they were harvesting them, even though everyone knows that peas take in radiation the most, as do all beans. And there are places out there with 40 curies or more. So he has no time for me at all. All the cooks and nurses have run off from the kindergartens. The kids are hungry. In order to take someone's appendix out, you need to take them in an ambulance to the next region, sixty kilometers on a road that's as bumpy as a washboard—all the surgeons have taken off. What car? What double furrows? He has no time for me.

So then I go to the military people. They were young guys, spending six months there. Now they're all very sick. They gave me an armored personnel carrier with a crew—no, wait, it was even better, it was an armored exploratory vehicle with a machine gun mounted on it. It's too bad I didn't get any photos of myself in it, on the armor. So, like I say, it was romantic. The ensign, who commanded the vehicle, was constantly radioing the base: "Eagle! Eagle! We're continuing our work." We're riding along, and these are our forests, our roads, but we're in an armored vehicle. The women are standing at their fences and crying—they haven't seen anything like this since the war. They're afraid another war has started.

According to the instructions, the tractors laying down the furrows were supposed to have driver's cabins that were hermetically sealed and protected. I saw the tractor, and the cabin was indeed hermetically sealed. But the tractor was sitting there and the driver was lying on the grass, taking a break. "Are you crazy? Haven't you been warned?" "But I put my sweatshirt over my head," he says. People didn't understand. They'd been frightened over and over again about a nuclear war, but not about Chernobyl.

It's such beautiful land out there. The old forests are still there, ancient forests. The winding little streams, the color of tea and clear as day. Green grass. People calling to each other through the forest. For them it was so natural, like waking up in the morning and walking out into your garden. And you're standing there knowing that it's all been poisoned.

We ran into an old lady.

"Children, tell me, can I drink milk from my cow?"

We look down at the ground, we have our orders—collect data, but don't interact with the locals.

Finally the driver speaks up.

"Grandma, how old are you?"

"Oh, more than eighty. Maybe more than that, my documents got burned during the war."

"Then drink all you want."

I feel worst of all for the people in the villages—they were innocent, like children, and they suffered. Farmers didn't invent Chernobyl, they had their own relations with nature, trusting relations, not predatory ones, just like they had one hundred years ago, and one thousand years ago. And they couldn't understand what had happened, they wanted to believe scientists, or any educated person, like they would a priest. But they were told: "Everything's fine. There's nothing to fear. Just wash your hands before eating." I understood, not right away, but after a few years, that we all took part in that crime, in that conspiracy. [*She is silent.*]

You have no idea how much of what was sent into the Zone as aid came out of it as contraband: coffee, canned beef, ham, oranges. It was taken out in crates, in vans. Because no one had those products anywhere. The local produce salesmen, the inspectors, all the minor and medium bureaucrats lived off this. People turned out to be worse than I thought. And me, too. I'm also worse. Now I know this about myself. [*Stops.*] Of course, I admit this, and for me that's already important. But, again, an example. In one kolkhoz there are, say, five villages. Three are "clean," two are "dirty." Between them there are maybe two to three kilometers. Two of them get "graveyard" money, the other three don't. Now, the "clean" village is building a livestock complex, and they need to get some clean feed. Where do they get it? The wind blows the dust from one field to the next, it's all one land. In order to build the complex, though, they need some papers signed, and the commission that signs them, I'm on the commission. Everyone knows we can't sign those papers. It's a crime. But in the end I found a justification for myself, just like everyone else. I thought: the problem of clean feed is not a problem for an environmental inspector.

Everyone found a justification for themselves, an explanation. I experimented on myself. And basically I found out that the frightening things in life happen quietly and naturally.

*Zoya Danilovna Bruk, environmental inspector*

## MONOLOGUE ABOUT ANSWERS

But haven't you noticed that we don't even talk about it among ourselves? In a few decades, in a hundred years, these will be mythic years.

I'm afraid of the rain. That's what Chernobyl is. I'm afraid of snow, of the forest. This isn't an abstraction, a mind game, but an actual human feeling. Chernobyl is in my home. It's in the most precious thing: my son, who was born in the spring of 1986. Now he's sick. Animals, even cockroaches, they know how much and when they should give birth. But people don't know how to do that, God didn't give us the power of foresight. A while ago in the papers it said that in Belarus alone, in 1993 there were 200,000 abortions. Because of Chernobyl. We all live with that fear now. Nature has sort of rolled up, waiting.

Zarathustra would have said: "Oh, my sorrow! Where has the time gone?"

I've thought about this a lot. I've searched for meaning in it. Chernobyl is the catastrophe of the Russian mind-set. Have you considered this? Of course I agree with those who write that it wasn't just the reactor that exploded, but an entire system of values. But this explanation isn't quite enough for me.

I'm a historian. I used to work on linguistics, the philosophy of language. We don't just think in language, but language thinks us. When I was eighteen, or maybe a little earlier, when I began to read samizdat and discovered Shalamov, Solzhenitsyn, I suddenly understood that my entire childhood, the childhood of my street, even though I grew up in a family that was part of the intelligentsia (my grandfather was a minister, my father a professor at the university at St. Petersburg), all of it was shot through with the language of the camps. For us as teenagers it was perfectly natural to call our fathers *pakhan*, our mothers *makhan*. "For every sneaky asshole there's a dick with a screwdriver"—I learned that saying when I was nine years old. I didn't know a single civilized word. Even our games, our sayings, our riddles were from the camps. Because the camps weren't a different world, which existed far away in the jails. It was right here. Akhmatova wrote, "Half the country was put away, half the country sits in jail." I think that this prison consciousness was inevitably going to collide with culture—with civilization, with the particle accelerator.

And of course we were raised with a particular Soviet form of paganism, which said that man was the crown of all creation, that it was his right to do anything with the world that he wanted. The Michurin formula: "We can't wait for favors from Mother Nature, we need to take them from her ourselves." It was an attempt to teach people the qualities that they didn't naturally possess. We had the psychology of oppressors. Now everyone talks about God. But why didn't they look for Him in the Gulag, or the jail cells of 1937, or at the Party meetings of 1948 when they started denouncing cosmopolitanism, or under Khrushchev when they were wrecking the old churches? The contemporary subtext of Russian religious belief is sly and false. They're bombing peaceful homes in Chechnya, they're destroying a small and proud people. That's the only way we know how to do it, with the sword—the Kalashnikov instead of the word. And we scrape out the incinerated Russian tank drivers with shovels—what's left of them. And nearby they're standing with candles in the church. For Christmas.

What now? We need to find out whether we're capable of the sort of total reconsideration of our entire history that the Germans and Japanese proved possible after the war. Do we have enough intellectual courage? People hardly talk about this. They talk about the market, about vouchers, about checks. Once again, we're just barely surviving. All our energy is directed toward that. But our souls have been abandoned.

So what is all this for? This book you're writing? The nights when I don't sleep? If our life is just a flick of the match? There might be a few answers to this. It's a primitive sort of fatalism. And there might be great answers to it, too. The Russian always needs to believe in something: in the railroad, in the frog (Bazarov), in Byzantium, in the atom. And now, in the market.

Bulgakov writes in *A Cabal of Hypocrites*: "I've sinned my whole life. I was an actor." This is a consciousness of the sinfulness of art, of the amoral nature of looking into another person's life. But maybe, like a small bit of disease, this could serve as inoculation against someone else's mistakes. Chernobyl is a theme worthy of Dostoevsky, an attempt to justify mankind.

Or maybe the moral is simpler than that: you should come into this world on your tiptoes, and stop at the entrance? Into this miraculous world . . .

*Aleksandr Revalskiy, historian*

MONOLOGUE ABOUT MEMORIES

I don't want to talk about this. I won't. I know just one thing: that I'll never be happy again.

He came back from there. He was there for a few years, like in a nightmare. "Nina," he said, "it's good we have two kids. They'll remain."

He told me stories. In the middle of a village there's a red puddle. The geese and ducks walk around it. The soldiers, they're just boys, are lying on the grass, their shirts off, their shoes off, tanning themselves. "Get up! Get up, you idiots, or you'll all die!" They say: ah, don't worry about it.

Death is already everywhere, but no one takes it seriously.

The evacuation: this old woman is on her knees with an icon in front of her old house. She says, "Boys, boys, I won't go. I won't leave this. You can take the little bit of money they gave me. For my house, for my cow, they gave it. But who'll pay me for my life? My life is a dark night. They killed my two sons during the war, they're in a little grave here. You call this a war? This is a war? There are white clouds in the sky, the apple trees are blooming. No one attacked us. No one's shooting. It's just us here. This is a war?" And no one can answer her: the colonel is standing there, the one who's running the evacuation, someone from the regional Party committee. The local bosses. No one knows that this is a war, and that it's called Chernobyl.

I never asked him myself. I understood him with my soul, we felt each other on a much deeper level. Our knowledge and our loneliness. That loneliness . . .

He knew he would die. That he was dying. He gave himself his word that he would live only through kindness and love. I worked two jobs, just one salary and his pension weren't enough. He said, "Let's sell the car. It's not new, but we'll get something from it. You'll stay home more, I'll look at you." He'd invite his friends over. His parents came to stay with us for a long time. He understood something. He understood something about life over there that he hadn't understood before. He found a different language.

"Nina," he'd say, "it's good that we have our two kids. They'll remain here."

I'd ask him, "Did you think of us? What did you think of there?"

"I saw a boy—he'd been born two months after the explosion. They named him Anton, but everyone called him Atomchik."

"You thought . . ."

"You felt sorry for everyone there. Even the flies, even the pigeons. Everyone should be able to live. The flies should be able to fly, and the wasps, the cockroaches should be able to crawl. You don't even want to hurt a cockroach."

"You . . ."

"The kids draw Chernobyl. The trees in the pictures grow upside-down. The water in the rivers is red or yellow. They'll draw it and then cry."

I want to understand . . . what? I don't know myself. [*Smiles.*] His friend proposed to me. He'd been in love with me long ago, back when we were in school. Then he married my friend, and then they got divorced. He made me an offer, "You'll live like a queen." He owns a store, he has a huge apartment in the city, he has a dacha. I thought and thought about it. But then one day he came in drunk: "You're not going to forget your hero, is that it? He went to Chernobyl, and I refused. I'm alive, and he's a memorial."

Ha-ha. I threw him out! I threw him right out! Sometimes I get strange thoughts, sometimes I think Chernobyl saved me, forced me to think. My soul expanded.

He told me about it and told me, and I remembered.

The clouds of dust, the tractors in the fields, the women with their pitchforks, the dosimeter clicking. Everything behind barbed wire. The Zone: no people, but time moves on anyway. The days are so long. Just like when you're a kid.

Entertainers came to visit them. Poets read their poems. Alla Pugacheva gave a concert in a field. "If you don't fall asleep, boys, I'll sing to you all night." She called them heroes.

Everyone called them heroes. [*Cries.*] It's impossible to suffer like this without any meaning. Without any of the old words. Even without the medal that they gave him. It's there at home. He gave it to our son. I know just one thing: I'll never be happy again.

*Nina Prohorovna Kovaleva, wife of a liquidator*

MONOLOGUE ABOUT LOVING PHYSICS

Ever since I was a youth I always wrote things down. When Stalin died, I wrote down everything happening in the streets, what people were saying. And I wrote about Chernobyl from the very first day, I knew time would pass and a lot of it would be forgotten and disappear forever. And that's how it was in fact. My friends were in the center of everything, they were nuclear physicists, but they forgot how they felt and what they spoke to me about. But I wrote everything down.

On that day, I came into work at the Institute for Nuclear Energy of the Belarus Academy of Sciences. I was the head of the laboratory there, the Institute was in the forest outside of town. It was wonderful weather! Spring. I opened the window. The air was fresh and clean, and I was surprised to see that for some reason the blue jays I'd been feeding all winter, hanging pieces of salami out the window for them, weren't around. Had they found a better meal somewhere else?

In the meantime there's panic at the reactor at the Institute: the dosimeters are showing heightened activity, readings are up 200 times on the air-cleaning filters. The force of the dose near the entrance is almost three milliroentgen per hour. That's very serious—that level is considered the highest allowable during work in radioactively dangerous zones for a maximum of six hours. The first theory was that a hermetic seal had broken on one of the heat-generating elements. We checked and it was fine. The second thought was that maybe the container from the radiochemical lab had been damaged in transit and had contaminated the whole territory? But that would be a spot on the sidewalk somewhere—try washing that out! So what's going on?

And at this point the internal radio announces that workers are advised not to leave the building. The area between our separate buildings grows deserted. Not a single person. It was frightening and strange.

The dosimetrists check my office—the desk is "glowing," my clothes are glowing, the walls are glowing. I get up, I don't even want to sit down in my chair. I wash my hair over the sink, check the dosimeter—it's gotten better. So is it possible that there's an emergency at our institute? Some leak? So how are we going to clean up the buses that take us around the city? We'll have to break our heads thinking of something. And I was very proud of our reactor, I'd studied every millimeter of it.

We call up the nearby Ignalinsk nuclear plant. Their instruments are also going crazy. They're also panicking. Then we call Chernobyl—and no one answers. By lunchtime we find out there's a radioactive cloud over all of Minsk. We determined that the activity was iodine in nature. That means the accident was at a reactor.

My first reaction was to call my wife, to warn her. But all our telephones at the Institute were bugged. Oh, that ancient fear, they'd been raising us on it for decades. But they don't know anything at home. My daughter is walking around with her friends after her music lesson at the conservatory. She's eating ice cream. Do I call? But that could lead to unpleasantness. They won't allow me to work on classified projects. But I can't take it, I pick up the phone.

"Listen to me carefully."

"What are you talking about?" my wife asks loudly.

"Not so loud. Close the windows, put all the food in plastic. Put on rubber gloves and wipe everything down with a wet cloth. Put the rag in a bag and throw it out. If there's laundry drying on the balcony, put it back in the wash."

"What happened?"

"Not so loud. Dissolve two drops of iodine in a glass of water. Wash your hair with it."

"What—" But I don't let her finish, I put down the phone. She should understand, she works at the Institute herself.

At 15:30, we learned that there'd been an accident at the Chernobyl reactor.

That evening on the way back to Minsk on the institute bus we rode for half an hour in silence, or talking of other things. Everyone was afraid to talk about what had happened. Everyone had his Party card in his pocket.

There was a wet rag in front of my apartment door—so my wife understood everything. I came in, threw off my jacket, and then my shirt, my pants, stripped down to my underwear. And suddenly this fury took hold of me. The hell with this secrecy! This fear! I took the city phone directory, and my daughter's address book and my wife's, and began calling everyone one by one. I'd say: I work at the Institute for Nuclear Physics. There is a radioactive cloud over Minsk. And then I'd tell them what they needed to do: wash their hair, close their windows, take the laundry off the balcony and wash it again, drink iodine, how to drink it correctly. People's reaction was: thank you. They didn't question me, they didn't get scared. I think they didn't believe me, or maybe they didn't understand the importance of what was taking place. No one became frightened. It was a surprising reaction.

That evening my friend calls. He was a nuclear physicist. And it was so careless! We lived with such belief! Only now can you see with what belief. He calls and says that, by the way, he's hoping to spend the May holidays at his in-laws' near Gomel. It's a stone's throw from Chernobyl. And he's bringing his little kids. "Great idea!" I yell at him. "You've lost your mind!" That's a tale of professionalism. And of our faith. But I yelled at him. He probably doesn't remember that I saved his children. [*Takes a break.*]

We—I mean all of us—we haven't forgotten Chernobyl. We never understood it. What do savages understand about lightning?

There's a moment in Ales Adamovich's book, when he's talking to Andrei Sakharov about the atom bomb. "Do you know," says Sakharov, the father of the hydrogen bomb, "how pleasantly the air smells of ozone after a nuclear explosion?" There is a lot of romance in those words. For me, for my generation—I'm sorry, I see by your reaction, you think I am celebrating something terrible, instead of human genius. But it's only now that nuclear energy has fallen so low and been shamed. But for my generation—in 1945, when they first dropped the atom bomb, I was seventeen years old. I loved science fiction, I dreamt of traveling to other planets, and I decided that nuclear energy would take us into the cosmos. I enrolled at the Moscow Energy Institute and learned that the most top-secret department was the nuclear energy department. In the fifties and sixties, nuclear physicists were the elite, they were the best and brightest. The humanities were pushed aside. Our teacher back in school would say, In three little coins there is enough energy to fuel an electrical power station. Your head could spin! I read the American Smith, who described how they invented the atomic bomb, tested it, what the explosions were like. In our world everything was a secret. The physicists got the high salaries, and the secrecy added to the romance. It was the cult of physics, the era of physics! Even when Chernobyl blew up, it took people a long time to part with that cult. They'd call up scientists, scientists would fly into Chernobyl on a special charter plane, but many of them didn't even take their shaving kits, they thought they'd be there just a few hours. Just a few hours, even though they knew that a reactor had blown up. They believed in their physics, they were of the generation that believed in it. But the era of physics ended at Chernobyl.

Your generation already sees the world differently. I recently read a passage in Konstantin Leontiev in which he writes that the results of man's physical-chemical experiments will lead a higher power to intervene in our earthly affairs. But for a person who was raised under Stalin, we couldn't imagine the possibility of some supernatural power. I only read the Bible afterwards. I married the same woman twice. I left and then came back—we met each other again in the same world. Life is a surprising thing! A mysterious thing! Now I believe. What do I believe in? I believe that the three-dimensional world has become crowded for mankind. Why is there such an interest in science fiction? Man is trying to tear himself away from the earth. He is trying to master different categories of time, different planets, not just this one. The apocalypse—nuclear winter—in Western literature this has already all been written, as if they were rehearsing it. They were preparing for the future. The explosion of a large number of nuclear warheads will result in enormous fires. The atmosphere will be saturated with smoke. Sunlight won't be able to reach the earth, and this will ignite a chain reaction—from cold to colder to colder still. This man-made version of the end of the world has been taught since the industrial revolution of the eighteenth century. But atom bombs won't disappear even after they destroy the last warhead. There will still be the knowledge of atom bombs.

You merely asked, but I keep arguing with you. We're having an argument between generations. Have you noticed? The history of the atom—it's not just a military secret and a curse. It's also our youth, our era, our religion. Fifty years have gone by, just fifty years. Now I also sometimes think that the world is being ruled by someone else, that we with our cannons and our spaceships are like children. But I haven't convinced myself of this yet.

Life is such a surprising thing! I loved physics and thought that I wouldn't ever do anything but physics. But now I want to write. I want to write, for example, about how man does not actually please science very much—he gets in the way of it.

Or about how a few physicists could change the world. About a new dictatorship of physics and math. A whole new life has opened up for me.

*Valentin Alekseevich Borisevich,*
*former head of the Laboratory of the Institute of Nuclear Energy*
*at the Belarussian Academy of Sciences*

## MONOLOGUE ABOUT EXPENSIVE SALAMI

In those first days, there were mixed feelings. I remember two: fear and insult. Everything had happened and there was no information: the government was silent, the doctors were silent. The regions waited for directions from the *oblast*, the *oblast* from Minsk, and Minsk from Moscow. It was a long, long chain, and at the end of it a few people made the decisions. We turned out to be defenseless. That was the main feeling in those days. Just a few people were deciding our fate, the fate of millions.

At the same time, a few people could kill us all. They weren't maniacs, and they weren't criminals. They were just ordinary workers at a nuclear power plant. When I understood that, I experienced a very strong shock. Chernobyl opened an abyss, something beyond Kolyma, Auschwitz, the Holocaust. A person with an ax and a bow, or a person with a grenade launcher and gas chambers, can't kill everyone. But with an atom . . .

I'm not a philosopher and I won't philosophize. Better to tell you what I remember. There was the panic in the first days: some people ran to the pharmacy and bought up all the iodine, others stopped going to the market, buying milk, meat, and especially lamb. Our family tried not to economize, we bought the most expensive salami, hoping that it would be made of good meat. Then we found out that it was the expensive salami that they mixed contaminated meat into, thinking, well, since it was expensive fewer people would buy it. We turned out to be defenseless. But you know that already. I want to tell about something else, about how we were a Soviet generation.

My friends—they were doctors and teachers, the local intelligentsia. We had our own circle, we'd get together at my house and drink coffee. Two old friends were there, one of them was a doctor, and they both had little kids.

"I'm leaving tomorrow to go live with my parents," says the doctor. "I'm taking the kids with me. I'd never forgive myself if they got sick."

"But the papers say that in a few days the situation will be stabilized," says the second. "Our troops are there. Helicopters, armored vehicles. They said so on the radio."

The first says, "You should take your kids too. Get them away from here! Hide them! This isn't a war. We can't even imagine what's happened."

And suddenly they took these tones with one another and it ended up in recriminations and accusations.

"What would happen if everyone behaved like you? Would we have won the war?"

"You're a traitor to your children! Where is your maternal instinct? Fanatic!"

And everyone's feeling there, including mine, was that my doctor friend was panicking. We needed to wait until someone told us, until they announced it. But she was a doctor, she knew more: "You can't even defend your own children! No one's threatening them? But you're afraid anyway!" How we hated her at that moment, she ruined the whole evening for us. The next day she left, and we all dressed up our kids and took them to the May Day demonstration. We could go or not go, as we pleased. No one forced us to go, or demanded that we go. But we thought it was our duty. Of course! At such a time, on such a day—everyone should be together. We ran along the streets, in the crowd.

All the secretaries of the regional Party committee were up on the tribunal, next to the first secretary. And his little daughter was there, standing so that everyone could see her. She was wearing a raincoat and a hat, even though it was sunny out, and he was wearing a military trench-coat. But they were there. I remember that. It's not just the land that's contaminated, but our minds. And for many years, too.

*From a letter from Lyudmila Dmitrievna Polenkaya,*
*village teacher, evacuated from the Chernobyl Zone*

## MONOLOGUE ABOUT FREEDOM AND THE
## DREAM OF AN ORDINARY DEATH

It was freedom. You felt like you were a free person there. That's not something you can understand, that's something only someone who's been in a war can understand. I've seen them, those guys—they get drunk, and they'll start talking about how they still miss it—the freedom, the flight. Not one step backward! Stalin's order. The special forces. But you shoot, you survive, you receive the 100 grams you're entitled to, a pouch of cheap tobacco. There are a thousand ways for you to die, to get blown to bits, but if you try hard enough, you can trick them—the devil, the senior officers, the combat, the one who's in that coffin with that wound, the very Almighty—you can trick them all and survive!

There's a loneliness to freedom. I know it, all the ones who were at the reactor know it. Like in a trench at the very front. Fear and freedom! You live for everything. That's not something you can understand, who live an ordinary life. Remember how they were always preparing us for war? But it turns out our minds weren't ready. I wasn't ready. Two military guys came to the factory, called me out. "Can you tell the difference between gasoline and diesel?" I say: "Where are you sending me?" "What do you mean where? As a volunteer for Chernobyl." My military specialty is rocket fuel. It's a secret specialty. They took me straight from the factory, just in a T-shirt, they didn't even let me go home. I said, "I need to tell my wife." "We'll tell her ourselves." There were about fifteen of us on the bus, reserve officers. I liked them. If we had to, we went, if it was needed, we worked, if they told us to go to the reactor, we got up on the roof of that reactor.

Near the evacuated villages they'd set up elevated guard posts, soldiers sat there with their rifles. There were barriers. Signs that said, "The side of the road is contaminated. Stopping or exiting is strictly forbidden." Gray trees, covered with decontamination-liquid. You start going crazy! Those first few days we were afraid to sit on the ground, on the grass, we didn't walk anywhere, we ran, if a car passed us, we'd put on a gas mask right away for the dust. After our shifts we'd sit in the tents. Ha! After a few months, it all seemed normal. It was just where you lived. We tore plums off the trees, caught fish, the pike there is incredible. And breams—we dried them to eat with beer. People probably told you about this already? We played soccer. We went swimming! Ha. [*Laughs again.*] We believed in fate, at bottom we're all fatalists, not pharmacists. We're not rational. That's the Slavic mind-set. I believed in my fate! Ha ha! Now I'm an invalid of the second category. I got sick right away. Radiation poisoning. I didn't even have a medical card at the polyclinic before I went. Ah, the hell with them. I'm not the only one. It was a mind-set.

I was a soldier, I closed other people's homes, went into them. It's a certain feeling . . . Land that you can't plant on—the cow butts its head against the gate, but it's closed and the house is locked. Its milk drips to the ground. There's a feeling you get! In the villages that hadn't been evacuated yet, the farmers would make vodka, they'd sell it to us. And we had lots of money: three times our salary from work, plus three times the normal daily military allowance. Later on we got an order: whoever drinks can stay a second term. So does vodka help or not? Well, at least psychologically it does. We believed that as much as we believed anything.

And the farmer's life flowed along very smoothly: they plant something, it grows, they harvest it, and the rest goes on without them. They don't have anything to do with the tsar, with the government—with space ships and nuclear power plants, with meetings in the capital. And they couldn't believe that they were now living in a different world, the world of Chernobyl. They hadn't gone anywhere. People died of shock. They took seeds with them, quietly, they took green tomatoes, wrapped them up. Glass cans would blow up, they'd put another back on the stove. What do you mean destroy, bury, turn everything into trash? But that's exactly what we did. We annulled their labor, the ancient meaning of their lives. We were their enemies.

I wanted to go to the reactor. "Don't worry," the others told me, "in your last month before demobilization they'll put you all on the roof." We were there six months. And, right on schedule, after five months of evacuating people, we were sent to the reactor. There were jokes, and also serious conversations, that we'd be sent to the roof. Well, after that maybe we'd live another five years. Seven. Maybe ten. But "five" is what people said most often, for some reason. Where'd they get that? And they said it quietly, without panicking. "Volunteers, forward march!" Our whole squadron—we stepped forward. Our commander had a monitor, he turned it on and it showed the roof of the reactor: pieces of graphite, melted bitumen. "See, boys, see those pieces there? You need to clean those up. And here, in this quadrant, is where you make the hole." You were supposed to be up there forty, fifty seconds, according to the instructions. But that was impossible. You needed a few minutes at the least. You had to get there and back, you had to run up and throw the stuff down—one guy would load the wheelbarrow, the others would throw the stuff into the hole there. You threw it, and went back, you didn't look down, that wasn't allowed. But guys looked down.

The papers wrote: "The air around the reactor is clean." We'd read it and laugh, then curse a little. The air was clean, but we got some serious dosage up there. They gave us some dosimeters. One was for five roentgen, and it went to the max right away. The other one was bigger, it was for 200 roentgen, and that too went off. Five years, they said, and you can't have kids. If you don't die after five years . . . [*Laughs.*] There were all kinds of jokes. But quietly, without panic. Five years . . . I've already lived ten. There! [*Laughs.*] They gave us decorations. I have two. With all the pictures: Marx, Engels, Lenin. Red flags.

One guy disappeared. We figured he'd run off, but then two days later we find him in the bushes, he'd hung himself. Everyone had this feeling, you understand . . . but then our political officer spoke, he said the guy had received a letter, his wife was cheating on him. Who knows? A week later we were demobilized. But we found him in the bushes.

We had a cook, he was so afraid that he didn't even live in the tent, he lived at the warehouse, he dug himself a little niche under the crates of butter and canned meat. He brought his mattress there and his pillow and lived underground. Then we got an order: gather a new crew and everyone to the roof. But everyone had already been. And they needed men! So they picked him up. He only went up on the roof once. He's a second-group invalid now. He calls me a lot. We keep in touch, we hold onto one another, to our memories, they'll live as long as we do. That's what you should write.

The papers are all lies. I didn't read anywhere about how we sewed ourselves protective gear, lead shirts, underwear. We had rubber robes with some lead in them. But we made lead underwear for ourselves. We made sure of that. In one village they showed us the two whorehouses. We were men who'd been torn away from our homes for six months, six months without women, it was an emergency situation. We all went there. The local girls would walk around anyway, even though they were crying, saying they'd all die soon. We had lead underwear, we wore it over our pants. Write that. We had good jokes, too. Here's one: An American robot is on the roof for five minutes, and then it breaks down. The Japanese robot is on the roof for five minutes, and then—breaks down. The Russian robot is up there two hours! Then a command comes in over the loudspeaker: "Private Ivanov! In two hours you're welcome to come down and have a cigarette break." Ha-ha! [*Laughs.*]

Before we went up on the roof, the commander gave us instructions, we were all standing as a unit, and a few of the guys protested: "We already went, we should have been sent home already." And as for me, my specialty was fuel, and they were sending me on the roof too. But I didn't say anything. I wanted to go. I didn't protest. The commander says: "Only volunteers go up on the roof. The rest can step aside, you'll have a talk with the military prosecutor." Well, those guys stood around, talked about it a little, and then agreed. If you took the oath, then you should do what you have to. I don't think any of us doubted that they'd put us in jail for insubordination. They'd put out a rumor that it would be two to three years. Meanwhile if a soldier got more than 25 roentgen, his superiors could be put in jail for poisoning their men. So no one got more than 25 roentgen. Everyone got less. You understand? But they were good kids. Two of them got sick, and this other one, he said, "I'll go." And he'd already been on the roof once that day. People respected him for it. And he got a reward: 500 rubles. Another guy was making a hole up top, and then it was time for him to stop. We're all waving at him: "Come down." But he's on his knees up there and he's whacking away. He needed to make a hole in that spot, so we could throw the debris down. He didn't get up until he'd made the hole. He got a reward—1000 rubles. You could buy two motorcycles for that back then. Now he's a first-group invalid. But for being afraid, you paid right away.

Demobilization. We got in cars. The whole way through the Zone we kept our sirens on. I look back on those days. I was close to something then. Something fantastical. I don't have the words to describe it. And the words, "gigantic," "fantastic," they just don't do it. I had this feeling . . . What? I haven't had that feeling again even in love.

*Aleksandr Kudryagin, liquidator*

## MONOLOGUE ABOUT THE SHADOW OF DEATH

You need facts and details of those days? Or just my story? For example, I was never a photographer, and there I started taking photographs, I happened to have a camera with me. I thought I was just doing it for myself. But now it's my profession. I couldn't rid myself of the new feeling that I had there. Does that make sense? [*As he talks he spreads photographs on the table, chair, windowsills: giant sunflowers the size of carriage wheels, a sparrow's nest in an empty village, a lonely village cemetery with a sign that says, "High radiation. Do not enter." A baby carriage in the yard of an abandoned house, the windows are boarded up, and in the carriage sits a crow, as if it's guarding its nest. The ancient sight of cranes over a field that's gone wild.*]

People ask me: "Why don't you take photos in color? In color!" But Chernobyl: literally it means *black event*. There are no other colors there. But my story? It's just commentaries to these [*points to the photographs*]. But all right. I'll try. Although it's all in here. [*Points again to the photographs.*]

At the time I was working at a factory, and also finishing my degree in history at the university by correspondence. At the factory I was a plumber, second class. They got us into a group and sent us off at emergency speed, like we were going to the front.

"Where are we going?"

"Where they tell you to go."

"What are we going to do?"

"What they tell you to do."

"But we're builders."

"Then you'll build. You'll build around."

We built support structures: laundries, warehouses, tents. I was assigned to unload cement. What sort of cement, and where from—no one checked that. They loaded it, we unloaded it. You spend a day shoveling that stuff and by the end only your teeth are showing. You're made of cement, of gray cement, and your special protective gear is too. You shake it off in the evening, and the next day you put it on again.

They held political discussions with us—they explained that we were heroes, accomplishing things, on the front line. It was all military language. But what's a bec? A curie? What's a milliroentgen? We ask our commander, he can't answer that, they didn't teach it at the military academy. Milli, micro, it's all Chinese to him. "What do you need to know for? Just do what you're ordered. Here you're soldiers." Yes, soldiers—but not convicts.

A commission came to visit. "Well," they told us, "everything here's fine. The background radiation is fine. Now, about four kilometers from here, that's bad, they're going to evacuate the people out of there. But here it's normal." They have a dosimetrist with them, he turns on the little box hanging over his shoulder and waves that long rod over our boots. And jumps to the side—it was an involuntary reaction, he couldn't help it.

But here's where the interesting part starts for you, for a writer. How long do you think we remembered that moment? Maybe a few days, at most. Russians just aren't about to think only of themselves, of their own lives, to think that way. Our politicians are incapable of thinking about the value of an individual life, but then we're not capable of it either. Does that make sense? We're just not built that way. We're made of different stuff. Of course we drank a lot in the Zone, we really drank. By nighttime there wasn't a sober guy around. Now, after the first couple of glasses some guys would get lonely, remember their wives, or their kids, or talk about their jobs. Curse out their bosses. But then later, after a bottle or two—the only thing we talked about was the fate of the country and the design of the universe. Gorbachev and Ligachev, Stalin. Are we a great empire, or not, will we defeat the Americans, or not? It was 1986—whose airplanes are better, whose space ships are more reliable? Well, okay, Chernobyl blew up, but we put the first man in space! Do you understand, we'd go on like that until we were hoarse, until morning. The fact that we don't have any dosimeters and they don't give us some sort of powder just in case? That we don't have washing machines so that we can launder our protective gear every day instead of twice a month? That was discussed last. In between. Damn it, that's just how we're built!

Vodka was more valuable than gold. And it was impossible to buy. Everything in the villages around us had been drunk: the vodka, the moonshine, the lotion, the nail polish, the aerosols. So picture us, with a three-liter bottle of moonshine, or instead a bottle of Shipr eau de cologne and we're having these endless conversations. There were teachers and engineers among us, and then the full international brigade: Russians, Belarussians, Kazakhs, Ukrainians. We had philosophical debates—about how we're the prisoners of materialism, and that limits us to the objects of this world, but Chernobyl is a portal to infinity. I remember discussions about the fate of Russian culture, its pull toward the tragic. You can't understand anything without the shadow of death. And only on the basis of Russian culture could you begin to make sense of the catastrophe. Only Russian culture was prepared for it. We'd been afraid of bombs, of mushroom clouds, but then it turned out like this; we know how a house burns from a match or a fuse, but this wasn't like anything else. We heard rumors that the flame at Chernobyl was unearthly, it wasn't even a flame, it was a light, a shining. Not blue, but more like the sky. And not smoke, either. The scientists had been gods, now they were fallen angels, demons even. The secrets of nature were hidden from them, and still are. I'm a Russian, from Bryanschin. We used to have an old man who sat on his stoop, the house is leaning over, it's going to fall apart soon, but he's talking about the fate of the world. Every little factory circle will have its Aristotle. And every beer stand. Meanwhile we're sitting right under the reactor. You can imagine how much philosophy there was.

Newspaper crews came to us, took photos. They'd have these invented scenes: they'd want to photograph the window of an abandoned house, and they'd put a violin in front of it; then they'd call the photo, "Chernobyl Symphony." But you didn't have to make anything up there. You wanted to just remember it: the globe in the schoolyard crushed by a tractor; laundry that's been hanging out on the balcony for a year and has turned black; abandoned military graves, the grass as tall as the soldier statue on it, and on the automatic weapon of the statue, a bird's nest. The door of a house has been broken down, it's all been robbed, but the curtains are still pulled back. People have left, but their photographs are still in the houses, like their souls.

There was nothing unimportant, nothing too small. I wanted to remember everything exactly and in detail: the time of day when I saw this, the color of the sky, my own feelings. Does that make sense? Mankind had abandoned these places forever. And we were the first to experience this "forever." You can't let go of a single tiny thing. The faces of the old farmers—they looked like icons. They were the ones who understood it least of all. They'd never left their yard, their land. They appeared on this earth, fell in love, raised bread with the sweat of their brow, continued their line. Waited for their grandchildren. And then, having lived this life, they left the land by going into the land, becoming the land. A Belarussian peasant hut! For us, city dwellers, the home is a machine for living in. For them it's an entire world, the cosmos. So you'd drive through these empty villages, and you so want to meet a human being. The churches have been robbed—you walk in and it smells of wax. You feel like praying.

I wanted to remember everything, so I started photographing it. That's my story. Not long ago we buried a friend of mine who'd been there. He died from cancer of the blood. We had a wake, and in the Slavic tradition we drank. And then the conversations began, until midnight. First about him, the deceased. But after that? Once more about the fate of the country and the design of the universe. Will Russian troops leave Chechnya or not? Will there be a second Caucasian war, or has it already started? Could Zhirinovsky become president? Will Yeltsin be re-elected? About the English royal family and Princess Diana. About the Russian monarchy. About Chernobyl, the different theories. Some say that aliens knew about the catastrophe and helped us out; others that it was an experiment, and soon kids with incredible talents will start to be born. Or maybe the Belarussians will disappear, like the Schythians. We're metaphysicians. We don't live on this earth, but in our dreams, in our conversations. Because you need to add something to this ordinary life, in order to understand it. Even when you're near death.

*Viktor Latun, photographer*


MONOLOGUE ABOUT A DAMAGED CHILD

The other day my daughter said to me: "Mom, if I give birth to a damaged child, I'm still going to love him." Can you imagine that? She's in the tenth grade, and she already has such thoughts. Her friends, too, they all think about it. Some acquaintances of ours recently gave birth to a son, their first. They're a young, handsome pair. And their boy has a mouth that stretches to his ears and no ears. I don't visit them like I used to, but my daughter doesn't mind, she looks in on them all the time. She wants to go there, maybe just to see, or maybe to try it on.

We could have left, but my husband and I thought about it and decided not to. We're afraid to. Here, we're all Chernobylites. We're not afraid of one another, and if someone gives you an apple or a cucumber from their garden, you take it and eat it, you don't hide it shamefully in your pocket, your purse, and then throw it out. We all share the same memories. We have the same fate. Anywhere else, we're foreign, we're lepers. Everyone is used to the words, "Chernobylites," "Chernobyl children," "Chernobyl refugees." But you don't know anything about us. You're afraid of us. You probably wouldn't let us out of here if you had your way, you'd put up a police cordon, that would calm you down. [*Stops.*] Don't try to tell me it's not like that. I lived through it. In those first days . . . I took my daughter and ran off to Minsk, to my sister. My own sister didn't let us into the her home, she had a little baby she was breast-feeding. Can you imagine that? We slept at the train station.

I had crazy thoughts. Where should we go? Maybe we should kill ourselves so as not to suffer? That was just in the first days. Everyone started imagining horrible diseases, unimaginable diseases. And I'm a doctor. I can only guess at what other people were thinking. Now I look at my kids: wherever they go, they'll feel like strangers. My daughter spent a summer at pioneer camp, the other kids were afraid to touch her. "She's a Chernobyl rabbit. She glows in the dark." They made her go into the yard at night so they could see if she was glowing.

People talk about the war, the war generation, they compare us to them. But those people were happy! They won the war! It gave them a very strong life-energy, as we say now, it gave them a really strong motivation to survive and keep going. They weren't afraid of anything, they wanted to live, learn, have kids. Whereas us? We're afraid of everything. We're afraid for our children, and for our grandchildren, who don't exist yet. They don't exist, and we're already afraid. People smile less, they sing less at holidays. The landscape changes, because instead of fields the forest rises up again, but the national character changes too. Everyone's depressed. It's a feeling of doom. Chernobyl is a metaphor, a symbol. And it's changed our everyday life, and our thinking.

Sometimes I think it'd be better if you didn't write about us. Then people wouldn't be so afraid. No one talks about cancer in the home of a person who's sick with it. And if someone is in jail with a life sentence, no one mentions that, either.

*Nadezhda Afanasyevna Burakova,*
*resident of the village of Khoyniki*

## MONOLOGUE ABOUT POLITICAL STRATEGY

I'm a product of my time. I'm a believing Communist. Now it's safe to curse us out. It's fashionable. All the Communists are criminals. Now we answer for everything, even the laws of physics.

I was the First Secretary of the Regional Committee of the Communist Party. In the papers they write that it was, you know, the Communists' fault: they built poor, cheap nuclear power plants, they tried to save money and didn't care about people's lives. People for them were just sand, the fertilizer of history. They can go to hell! That's where! It's the cursed questions: what to do and who to blame? These are questions that don't go away. Everyone is impatient, they want revenge, they want blood. Well they can go to hell!

Others keep quiet, but I'll tell you. You write—well not you personally, but the papers write that the Communists fooled the people, hid the truth from them. But we had to. We got telegrams from the Central Committee, from the Regional Committee, they told us: you must prevent a panic. And it's true, a panic is a frightening thing. Only during the war did they pay so much attention to news from the front as they did then to the news from Chernobyl. There was fear, and there were rumors. People weren't killed by the radiation, but by the events. We had to prevent a panic.

You can't say that we covered everything up right away, we didn't even know the extent of what was happening. We were directed by the highest political strategy. But if you put aside the emotions, and the politics, you have to admit that no one believed in what had happened. Even the scientists couldn't believe it! Nothing like it had ever happened, not just here but anywhere in the world. The scientists who were there at the plant studied the situation and made immediate decisions. I recently watched the program, "Moment of Truth," where they interviewed Aleksandr Yakovlev, a member of the Politburo, the main ideologue of the Party under Gorbachev. What did he remember? They also, the ones up top, didn't really see the whole picture. At a meeting of the Politburo one of the generals explained: "What's radiation? At the testing grounds, after an atomic blast, they drink a bottle of wine and that's that. It'll be fine." They talked about Chernobyl like it was an accident, an ordinary accident.

What if I'd declared then that people shouldn't go outside? They would have said: "You want to disrupt May Day?" It was a political matter. They'd have asked for my Party card! [*Calms down a little.*] Here's something that happened, I think, it wasn't just a story. The chairman of the Government Commission, Scherbin, comes to the plant in the first days after the explosion and demands that they take him to the reactor. They say, No, there's chunks of graphite, insane radiation, high temperature, the laws of physics, you can't go. "What laws of physics! I need to see everything with my own eyes. I need to give a report tonight to the Politburo." It was the military way of dealing with things. They didn't know any other way. They didn't understand that there really is such a thing as physics. There is a chain reaction. And no orders or government resolutions can change that chain reaction. The world is built on physics, not on the ideas of Marx. But if I'd said that then? Tried to call off the May Day parade? [*Gets upset again.*] In the papers they write that the people were out in the street while we were in underground bunkers? I stood on the tribune for two hours in that sun, without a hat, without a raincoat. And on May 9, the Day of Victory, I walked with the veterans. They played the harmonica, people danced, drank. We were all part of that system. We believed! We believed in the high ideals, in victory! We'll defeat Chernobyl! We read about the heroic battle to put down the reactor that had gotten out from under man's control. A Russian without a high ideal? Without a great dream? That's also scary.

But that's what's happening now. Everything's falling apart. No government. Stalin. *Gulag Archipelago.* They pronounced a verdict on the past, on our whole life. But think of the great films! The happy songs! Explain those to me. Why don't we have such films anymore? And such songs? Man needs to be uplifted, inspired. He needs ideals. Only then will you have a strong nation. Shining ideals—we had those!

In the papers—on the radio and television they were yelling, Truth! Truth! At all the meetings they demanded: truth! Well, it's bad, it's very bad. We're all going to die! But who needs that kind of truth? When the mob tore into the convent and demanded the execution of Robespierre, were they right? You can't listen to the mob, you can't become the mob. Look around. What's happening now? [*Silent.*] If I'm a criminal, why is my granddaughter, my little child, also sick. My daughter had her that spring, she brought her to us in Slavgorod in diapers. In a baby carriage. It was just a few weeks after the explosion at the plant. There were helicopters flying, military vehicles on the roads. My wife said: "They should go to our relatives. They need to get out of here." I was the First Secretary of the Regional Committee of the Party. I said absolutely not. "What will people think if I take my daughter with her baby out of here? Their children have to stay." Those who tried to leave, to save their own skins, I'd call them into the regional committee. "Are you a Communist or not?" It was a test for people. If I'm a criminal, then why was I killing my own grandchild? [*Goes on for some time but it is impossible to understand what he's saying.*]

You asked me to tell you about the first days. In the Ukraine they had an alarm, but here in Belarus it's all calm. It's planting season. I didn't hide, I didn't sit in my office, I ran around the fields and meadows. People were planting and digging. Everyone forgets that before Chernobyl everyone called the atom the "peaceful worker," everyone was proud to live in the atomic age. I don't remember any fear of the atom.

After all, what is the First Secretary of the Regional Committee of the Party? An ordinary person with an ordinary college diploma, most often in engineering or agronomics. Some of us had also gone to the Higher Party School. I knew as much about the atom as they'd had a chance to tell me during courses on civil defense. I didn't hear a word about cesium in milk, or about strontium. We shipped milk with cesium to the milk plants. We gave meat to the meat plants. We harvested wheat. We carried out the plan. I didn't beat it out of them. No one called off the plans for us.

Here's a story about how we were then. During those first days, people felt fear, but also they were excited. I'm a person who lacks the instinct of self-preservation. [*Considers this.*] But people have a strong sense of duty. I had dozens of letters on my desk by people asking to be sent to Chernobyl. Volunteers. No matter what they write now, there was such a thing as a Soviet person, with a Soviet character. No matter what they write about it now.

Scientists came, they'd argue until they were yelling, until they were hoarse. I came up to one of them: "Are our kids playing in radioactive sand?" And he says: "They're alarmists! Dilettantes! What do you know about radiation? I'm a nuclear physicist. I've seen atomic explosions. Twenty minutes later I can drive up to the epicenter in a truck, on the melted earth. So why are you all raising a panic?" I believed them. I called people into my office. "Brothers! If I run away, and you run away, what will people think of us? They'll say the Communists are deserters." If I couldn't convince them with words, with emotions, I did it in other ways. "Are you a patriot or not? If not, then put your Party card on the table. Throw it down!" Some did.

I began to suspect something, though. We had a contract with the Institute for Nuclear Physics to run tests on the earth samples we sent them. They took grass, and then layers of black earth and brought them back to Minsk. They ran their analyses. And then they called me: "Please organize a transport to bring your soil back from here." "Are you kidding? It's four hundred kilometers to Minsk." The receiver almost fell out of my hand. "Take the earth back here?" They answer: "No, we're not kidding. According to our instructions these samples are to be buried in special containers, in an underground concrete-and-metal bunker. But we get them from all over Belarus. We filled up all the space we had in a month." Do you hear that? And we harvest and plant on this land. Our kids play on it. We're supposed to fulfill the plans for milk and meat. We make vodka out of the wheat. Then apples, pears, cherries go for juices.

The evacuation—if anyone had seen it from above, they would have thought that Word War III had begun. They evacuate one village, and then they tell the other village: your evacuation is in one week. The whole week they stack straw, mow the grass, dig in their gardens, chop wood. Just go about their lives, not understanding what's happening. And then a week later they all get taken out in military vehicles. For me there were meetings, business trips, tension, sleepless nights. So much was going on. I remember a guy standing next to the City Committee of the Party in Minsk with a sign: "Give the people iodine." It's hot, but he's wearing a raincoat.

You forget—people used to think nuclear power plants were our future. I spoke many times, I propagandized. I went to one nuclear plant—it was quiet, very celebratory, clean. In the corner a red flag, the slogan, "Winner of the Socialist Competition." Our future.

I'm a product of my time. I'm not a criminal.

*Vladimir Matveevich Ivanov,*
*former First Secretary of the Stavgorod Regional Party Committee*

## MONOLOGUE BY A DEFENDER OF THE
## SOVIET GOVERNMENT

What're you writing there? Who gave you permission? And taking pictures. Put that away. Put the camera away or I'll break it. How do you like that, coming here, writing things down. We live here. And you come around putting ideas in people's heads. Saying things. You're talking about the wrong things. There's no order now! How do you like that, coming around with their microphones.

That's right, I'm defending the Soviet government! Our government! The people's government! Under the Soviets we were strong, everyone was afraid of us. The whole world was watching us! Some were scared shitless, some were jealous. Fuck! And now what? What do we have now under democracy? They send us Snickers and some old margarine, old jeans, like we're savages who just climbed out of the trees. The palm trees. We had a great empire! And now how do you like that, coming around here. A great empire! Fuck!

Until Gorbachev came along. A devil with a birthmark. Gorbie. Gorbie was working for them, for the CIA. What are you trying to tell me, huh? Coming around here. They're the ones who blew up Chernobyl. The CIA and the democrats. I read it in the papers. If Chernobyl hadn't blown up, the empire wouldn't have collapsed. A great empire! Fuck! Now they're coming around here. You could buy a loaf of bread for twenty kopeks under the Communists, now it costs two thousand rubles. What did the democrats do? They sold everything off! Ransomed it! Our grandchildren won't have shit.

I'm not a drunk, I'm a Communist! They were for us, for the simple people. Don't feed me fairy tales about democracy and freedom. Fuck! This free person dies, there's nothing to bury him with. Old lady died not long ago. She was lonely, no children. She lay in her house for two days, in her old house shirt, under her icons. They couldn't afford a coffin. She was a hard worker, a Stakhanovite. We refused to go out to the fields for two days. We had meetings. Fuck! We met together until the chairman of the kolkhoz came out and promised that every person who died on the kolkhoz got a free coffin, and then a calf or a pig and two cases of vodka for the wake. Under the democrats—two cases of vodka—free! Half a bottle is medicine, for us, from the radiation, but a whole bottle a man is a party.

Why aren't you writing this down? What I'm saying? You only write down what you want to hear. Giving people ideas. Saying things. You need political capital, is that it? Stuff your pockets with dollars? We live here, we survive here. No one's guilty! Show me the guilty ones! I'm for the Communists. They'll come back and they'll find the guilty ones. Fuck! Coming around here, writing things down.

*No name given*

## MONOLOGUE ABOUT INSTRUCTIONS

I have a lot of material, I've been collecting it now for seven years—newspaper clippings, my own comments. I have numbers. I'll give you all of it. I'll never leave this subject now, but I can't write it myself. I can fight—organize demonstrations, pickets, get medicine, visit sick kids—but I can't write about it. You should, though. I just have so many feelings, I'll never be able to deal with them, they'll paralyze me. Chernobyl already has its own obsessives, its own writers. But I don't want to become one of the people who exploits this subject.

But if I were to write honestly? [*Thinks.*] That warm April rain. Seven years now I've thought about that rain. The raindrops rolled up like quicksilver. They say that radiation is colorless, but the puddles that day were green and bright yellow. My neighbor told me in a whisper that Radio Freedom had reported an accident at the Chernobyl nuclear power plant. I didn't pay her any mind. I was absolutely certain that if anything serious happened, they'd tell us. They have all kinds of special equipment—special warning signals, bomb shelters—they'll warn us. We were sure of it! We'd all taken civil defense courses, I'd even taught them. But that evening another neighbor brought me some powders. A relative had given them to her and explained how to take them, he worked at the Institute for Nuclear Physics, but he made her promise to keep quiet. As quiet as a fish! As a rock! He was especially worried about talking on the phone.

My nephew was living with me then, he was little. And me? I still didn't believe it. I don't think any of us drank those powders. We were very trusting—not just the older generation, but the younger one, too.

I'm remembering those first impressions, those first rumors, and I go from this time to that, from this condition to that one. It's difficult, from here—as a writer, I've thought about this, how it's as if there are two people inside me, the pre-Chernobyl me and the post-Chernobyl one. And it's very hard now to recall with any certainty what that "pre-" me was like. My vision has changed since then.

I started going to the Zone from the very first days. I remember stopping in some village, and I was shocked by how silent it was. No birds, nothing. You're walking down the street and there's—nothing. Silence. I mean, all right, the houses are empty, the people have all left, but all around everything's just shut down, there's not a single bird.

We got to the village of Chudyany—they have 149 curies. Then the village of Malinovka—59 curies. The people were absorbing doses that were a hundred times those of soldiers who patrol areas where nuclear testing takes place. Nuclear testing grounds. A thousand times! The dosimeter is shaking, it's gone to its limit, but the kolkhoz offices have signs up from the regional radiologists saying that it is all right to eat salad: lettuce, onions, tomatoes, cucumbers—all of it. Everything's growing and everyone's eating. What do those radiologists say now? And the regional Party secretaries? How do they justify themselves?

In the villages we met a lot of drunk people. They were walking around on benders, even the women, especially the milkmaids.

In one village we visited a kindergarten. The kids are running around, playing in the sandbox. The director told us they got new sand every month. It was brought in from somewhere. You can probably guess where they brought it from. The kids are all sad. We make jokes, they don't smile. Their teacher says, "Don't even try. Our kids don't smile. And when they're sleeping they cry." We met a woman on the street who'd just given birth. "Who let you give birth here?" I said. "It's 59 curies outside." "The doctor-radiologist came. She said I shouldn't dry the baby clothes outside." They tried to convince people to stay. Even when they evacuated a village, they still brought people back in to do the farming, harvest the potatoes.

What do they say now, the secretaries of the regional committees? How do they justify it? Whose fault do they say it is?

I've kept a lot of instructions—top-secret instructions, I'll give them all to you, you need to write an honest book. There are instructions for what to do with contaminated chickens. You were to wear protective gear just as you would if handling any radioactive materials: rubber gloves and rubber robes, boots, and so on. If there are a certain number of curies, you need to boil it in salt water, pour the water down the toilet, and use the meat for pate or salami. If there's more curie than that, then you put it into bone flour, for livestock feed. That's how they fulfilled the plans for meat. They were sold cheaply from the contaminated areas—into clean areas. The drivers who were taking them told me that the calves were strange, they had fur down to the ground, and they were so hungry they'd eat anything—rags, paper. They were easy to feed. They'd sell them to the kolkhoz, but if any of the drivers wanted one, they could take it for themselves, into their own farm. This was criminal! Criminal!

We met a small truck on the road. It was going so slowly, like it was driving to a funeral and there was a body in back. We stopped the car, I thought the driver was drunk, and there was a young guy at the wheel. "Are you all right?" I said. "Yes, I'm just carrying contaminated earth." In that heat! With all the dust! "Are you crazy? You still have to get married, raise kids!" "Where else am I going to get fifty rubles for a single trip?" Fifty rubles back then could get you a nice suit. And people talked more about the rubles than the radiation. They got these tiny bonuses. Or tiny anyway compared with the value of a human life.

It was funny and tragic at the same time.

*Irina Kiseleva, journalist*

## MONOLOGUE ABOUT THE LIMITLESS POWER
## ONE PERSON CAN HAVE OVER ANOTHER

I'm not a literary person, I'm a physicist, so I'm going to give you the facts, only facts.

Someone's eventually going to have to answer for Chernobyl. The time will come when they'll have to answer for it, just like for 1937. It might be in fifty years, everyone might be old, they might be dead. They're criminals! [*Quiet.*] We need to leave facts behind us. They'll need them.

On that day, April 26, I was in Moscow on business. That's where I learned about the accident.

I called Nikolai Slyunkov, the General Secretary of the Central Committee Belarussian Communist Party, in Minsk. I called once, twice, three times, but they wouldn't connect me. I reached his assistant, he knew me well.

"I'm calling from Moscow. Get me Slyunkov, I have information he needs to hear right away. Emergency information."

I'm calling over a government line, but they're already blocking things. As soon as you start talking about the accident, the line goes dead. So they're listening, obviously! I hope it's clear who's listening—the appropriate agency. The government within the government. And this is despite the fact that I'm calling the First Secretary of the Central Committee. And me? I'm the director of the Institute for Nuclear Energy at the Belarussian Academy of Science. Professor, member-correspondent of the Academy. But even I was blocked.

It took me about two hours to finally reach Slyunkov. I tell him: "It's a serious accident. According to my calculations"—and I'd had a chance by then to talk with some people in Moscow and figure some things out—"the radioactive cloud is moving toward us, toward Belarus. We need to immediately perform an iodine prophylaxis of the population and evacuate everyone near the station. No man or animal should be within 100 kilometers of the place."

"I've already received reports," says Slyunkov. "There was a fire, but they've put it out."

I can't hold it in. "That's a lie! It's a blatant lie! Any physicist will tell you that graphite burns at something like five tons per hour. Think of how long it's going to burn!"

I get on the first train to Minsk. I don't sleep the whole night there. In the morning I'm home. I measure my son's thyroid—that was the ideal dosimeter then—it's at 180 micro-roentgen per hour. He needed potassium iodine. This was ordinary iodine. A child needed two to three drops in half a glass of solvent, an adult needed three to four. The reactor burned for ten days, and this should have been done for ten days. But no one listened to us! No one listened to the scientists and the doctors. They pulled science and medicine into politics. Of course they did! We shouldn't forget the background to this, what we were like then, what we were like ten years ago. The KGB was working, making secret searches. "Western voices" were being shut out. There were a thousand taboos, Party and military secrets. And in addition everyone was raised to think that the peaceful Soviet atom was as safe as peat or coal. We were people chained by fear and prejudices. We had the superstition of our faith.

But, all right, the facts. That next day, April 27, I decided to go to the Gomel region on the border with the Ukraine. I went to the major towns—Bragin, Khoyniki, Narovlya, they're all just twenty, thirty kilometers from the station. I needed more information, I took all the instruments to measure the background radiation. The background was this: in Bragin, 30,000 micro-roentgen per hour; in Narovlya: 28,000. But people are out in the fields sowing, mowing, getting ready for

Easter. They're coloring eggs, baking Easter cakes. They say, What radiation? What's that? We haven't received any orders. The only thing we're getting from above is: how is the harvest, what's the pace now? They look at me like I'm crazy. "What do you mean, Professor?" Roentgen, micro-roentgen—this is the language of someone from another planet.

So we come back to Minsk. Everyone's out on the streets, people are selling pies, ice cream, sandwiches, pastries. And overhead there's a radioactive cloud.

On April 29—I remember everything exactly, by the dates—at 8 A.M. I was already sitting in Slyunkov's reception area. I'm trying to get in, and trying. They don't let me in. I sit there until half past five. At half past five, a famous poet walks out of Slyunkov's office. I know him. He says to me, "Comrade Slyunkov and I discussed Belarussian culture."

I explode: "There won't be any Belarussian culture or anyone to read your books, if we don't evacuate everyone from Chernobyl right away! If we don't save them!"

"What do you mean? They've already put it out."

I finally get in to see Slyunkov. I tell him what I saw the day before. We have to save these people! In the Ukraine—I'd called there—they're already evacuating.

"Why are your men" (from the Institute) "running around town with their dosimeters, scaring everyone? I've already consulted with Moscow, with Professor Ilyin, Chairman of the Soviet Radiological Protection Board. He says everything's normal. And there's a Government commission at the station, and the prosecutor's office is there. We've thrown the army, all our military equipment, into the breach."

We already had thousands of tons of cesium, iodine, lead, circonium, cadmium, berillium, borium, an unknown amount of plutonium (the uranium-graphite reactors of the Chernobyl variety also produced weapons-grade plutonium, for nuclear bombs)—450 types of radionuclides in all. It was the equivalent of 350 atomic bombs dropped on Hiroshima. They needed to talk about physics, about the laws of physics, but instead they talked about enemies, about looking for enemies.

Sooner or later, someone will have to answer for this. "You're going to say that you're a tractor specialist," I said to Slyunkov, he'd been a director of a tractor factory, "and that you didn't understand what radiation could do, but I'm a physicist, I know what the consequences are." But from his point of view, what was this? Some professor, a bunch of physicists, were going to tell the Central Committee what to do? No, they weren't a gang of criminals. It was more like a conspiracy of ignorance and obedience. The principle of their lives, the one thing the Party machine had taught them, was never to stick their necks out. Better to keep everyone happy. Slyunkov was just then being called to Moscow for a promotion. He was so close! I'd bet there'd been a call from the Kremlin, right from Gorbachev, saying, you know, I hope you Belarussians can keep from starting a panic, the West is already making all kinds of noises. And of course if you didn't please your higher-ups, you didn't get that promotion, that trip abroad, that dacha. If we were still a closed system, behind the Iron Curtain, people would still be living next to the station. They'd have covered it up! Remember—Kytrym, Semipalatinsk—we're still Stalin's country, you know.

In the civil defense instructions we had then, you were supposed to carry out an iodine prophylaxis for the entire population if there was the threat of a nuclear accident or nuclear attack. That was in the event of a *threat*. Here we had three thousand micro-roentgen per hour. But they're worried about their authority, not about the people. It was a country of authority, not people. The State always came first, and the value of a human life was zero. Because they might have found ways—without any announcements, without any panic. They could simply have introduced iodine into the freshwater reservoirs, or added it to the milk. The city had 700 kilograms of iodine concentrate for that very purpose—but it just stayed where it was. People feared their superiors more than they feared the atom. Everyone was waiting for the order, for a call, but no one did anything himself.

I carried a dosimeter in my briefcase. Why? Because they'd stopped letting me in to see the important people, they were sick of me. So I'd take my dosimeter along and put it up to the thyroids of the secretaries or the personal chauffeurs sitting in the reception rooms. They'd get scared, and sometimes that would help, they'd let me through. And then people would say to me: "Professor, why are you going around scaring everyone? Do you think you're the only one worried about the Belarussian people? And, anyway, people have to die of something, whether it's smoking, or an auto accident, or suicide." They laughed at the Ukrainians. They were on their knees at the Kremlin asking for more money, medicine, radiation-measuring equipment (there wasn't enough). Meanwhile our man, Slyunkov, had taken fifteen minutes to lay out the situation. "Everything's fine. We'll handle it ourselves." They praised him: "That's how it's done, our Belarussian brothers!" How many lives did that bit of praise cost?

I have information indicating that the bosses were taking iodine. When my colleagues at the Institute gave them checkups, their thyroids were clean. Without iodine that's impossible. And they quietly got their kids out of there, too, just in case. And when they went into the area themselves they had gas masks and special robes—the very things everyone else lacked. And it's no secret that they had a special herd near Minsk—every cow had a number and was watched over. They had special lands, special seedbeds, special oversight. And the most disgusting thing—no one's ever answered for it.

So they stopped receiving me in their offices. I bombarded them with letters instead. Official reports. I sent around maps, figures, to the entire chain of command. Four folders with 250 pages in each, filled with facts, just facts. I made two copies of everything, just in case—one was in my office at the Institute, the other I hid at home. My wife hid it. Why did I make the copies? That's the sort of country we live in. Now, I always locked up my office myself, but I came back from one trip, the folders were gone. But I grew up in the Ukraine, my grandfathers were Cossacks, I have a Cossack character. I kept writing. I kept speaking out. You need to save people! They need to be evacuated immediately! We kept making trips out there. Our institute was the first to put together a map of the contaminated areas. The whole south was red.

This is already history—the history of a crime.

They took away all the Institute's radiation-measuring equipment. They just confiscated it, without any explanations. I began receiving threatening phone calls at home. "Quit scaring people, Professor. You'll end up in a bad place. Want to know how bad? We can tell you all about it." There was pressure on scientists at the Institute, scare tactics.

I wrote to Moscow.

After that, Platonov, the president of the Academy of Sciences, called me in. "The Belarussian people will remember you someday, you've done a lot for them, but you shouldn't have written to Moscow. That was very bad. They're demanding that I relieve you of your post. Why did you write? Don't you understand who you're going up against?"

Well, I have maps and figures. What do they have? They can put me in a mental hospital. They threatened to. And they could make sure I had a car accident—they warned me about that, also. They could drag me into court for anti-Soviet propaganda. Or for a box of nails missing from the Institute's inventory.

So they dragged me into court.

And they got what they wanted. I had a heart attack. [*Silent.*]

I wrote everything down. It's all in the folder. It's facts, only facts.

We're checking the kids in the villages, the boys and girls. They have fifteen hundred, two thousand, three thousand micro-roentgen. More than three thousand. These girls—they can't give birth to anyone. They have genetic mutations. The tractor is plowing. I ask the Party worker who's with us: "Does the tractor driver at least wear a gas mask?"

"No, they don't wear them."

"What, you didn't get them?"

"Oh, we got plenty! We have enough to last until the year 2000. We just don't give them out, otherwise there'd be a panic. Everyone would run off, they'd leave."

"How can you do that?"

"Easy for you to say, Professor. If you lose your job, you'll find another one. Where am I going to go?"

What power! This limitless power that one person could have over another. This isn't a trick or lie anymore, it's just a war against the innocent.

Like when we're driving along the Pripyat. People have set up tents, they're camping out with their families. They're swimming, tanning. They don't know that for several weeks now they've been swimming and tanning underneath a nuclear cloud. Talking to them was strictly forbidden. But I see children, and I go over and start explaining. They don't believe me. "How come the radio and television haven't said anything?" My escort—someone from the local Party office was always with us—he doesn't say anything. I can follow his thoughts on his face: should he report this or not? But at the same time he feels for these people! He's a normal guy, after all. But I don't know what's going to win out when we get back. Will he report it or not? Everyone made his own choice. [*Extended silence.*]

What do we do with this truth now? How do we handle it? If it blows up again, the same thing will happen again. We're still Stalin's country. We're still Stalin's people.

*Vasily Borisovich Nesterenko, former director of the*
*Institute for Nuclear Energy at the Belarussian Academy of Sciences*

MONOLOGUE ABOUT WHY WE LOVE CHERNOBYL

It was 1986—what were we like, then? How did this technological version of the end of the world find us? We were the local intelligentsia, we had our own circle. We lived our own lives, staying far away from everything around us. It was the form our protest took. We had our own rules: we didn't read the newspaper *Pravda*, but we handed the magazine *Ogonyok* from hand to hand. They had just loosened the reins a little, and we were drinking it all in. We read Solzhenitsyn, Shalamov, went to each other's houses, had endless talks in the kitchen. We wanted something more from life. What? Somewhere there were movie actors—Catherine Deneuve—in a beret. We wanted freedom. Some people from our circle fell apart, drank themselves away, some people made a career for themselves, joined the Party. No one thought this regime could crumble. And if that's how it was, we thought, if it was going to last forever, then the hell with everyone. We'll just live here in our own little world.

Chernobyl happened, and at first we had the same kind of reaction. What's it to us? Let the authorities worry about it. That's their duty—Chernobyl. And it's far away. We didn't even look on the map. We didn't even need to know the truth, at that point.

But when they put labels on the milk that said, "For children," and "For adults"—that was a different story. That was a bit closer to home. All right, I'm not a member of the Party, but I still live here. And we became afraid. "Why are the radish leaves this year so much like beet leaves?" You turned on the television, they were saying, "Don't listen to the provocations of the West!" and that's when you knew for sure.

And the May Day parade? No one forced us to go—no one forced me to go there. We all had a choice and we failed to make it. I don't remember a more crowded, cheerful May Day parade. Everyone was worried, they wanted to become part of the herd—to be with others. People wanted to curse someone, the authorities, the government, the Communists. Now I think back, looking for the break. Where was it? But it was before that. We didn't even want to know the truth. We just wanted to know if we should eat the radish.

I was an engineer at the Khimvolokno factory. There was a group of East German specialists there at the time, putting in new equipment. I saw how other people, from another culture, behaved. When they learned about the accident, they immediately demanded medical attention, dosimeters, and a controlled food supply. They listened to the German radio programs and knew what to do. Of course, they were denied all their requests. So they immediately packed their bags and got ready to leave. Buy us tickets! Send us home! If you can't provide for our safety, we're leaving. They protested, sent telegrams to their government. They were fighting for their wives, their kids, they had come here with their families, they were fighting for their lives! And us? How did we behave? Oh, those Germans, they're all so nicely taken care of, they're all so arrogant—they're hysterical! They're cowards! They're measuring the radiation in the borsch, in the ground meat. What a joke! Now, our men, they're real men. Real Russian men. Desperate men. They're fighting the reactor. They're not worried about their lives. They get up on that melting roof with their bare hands, in their canvas gloves (we'd already seen this on television). And our kids go with their flags to the demonstrations. As do the war veterans, the old guard. [*Thinks.*] But that's also a form of barbarism, the absence of fear for oneself. We always say "we," and never "I." "We'll show them Soviet heroism," "we'll show them what the Soviet character is made of." We'll show the whole world! But this is me, this is I. I don't want to die. I'm afraid.

It's interesting to watch oneself from here, watch one's feelings. How did they develop and change? I've noticed that I pay more attention to the world around me. After Chernobyl, that's a natural reaction. We're beginning to learn to say "I." I don't want to die! I'm afraid.

That great empire crumbled and fell apart. First, Afghanistan, then Chernobyl. When it fell apart, we found ourselves all alone. I'm afraid to say it, but we love Chernobyl. It's become the meaning of our lives. The meaning of our suffering. Like a war. The world found out about our existence after Chernobyl. It was our window to Europe. We're its victims, but also its priests. I'm afraid to say it, but there it is.

Now it's my work. I go there, and look. In the Zone the people live in fear still, in their crumbling cottages. They want Communism again. At all the elections they vote for the firm hand, they dream of Stalin's times, military times. And they live there in a military situation: police posts, people in uniform, a pass system, rationing, and bureaucrats distributing the humanitarian aid. It says on the boxes in German and Russian: "Cannot be sold or exchanged." But it's sold and exchanged right next door, in every little kiosk.

And it's like a game, like a show. I'm with a caravan of humanitarian aid and some foreigners who've brought it, whether in the name of Christ or something else. And outside, in the puddles and the mud in their coats and mittens is my tribe. In their cheap boots. "We don't need anything," their eyes seem to be saying, "it's all going to get stolen anyway." But also the wish to grab a bit of something, a box or crate, something from abroad. We know where all the old ladies live by now. And suddenly I have this outrageous, disgusting wish. "I'll show you something!" I say. "You'll never see this in Africa! You won't see it anywhere. 200 curie, 300 curie." I've noticed how the old ladies have changed, too—some of them are real actresses. They know their monologues by heart, and they cry in all the right spots. When the first foreigners came, the grandmas wouldn't say anything, they'd just stand there crying. Now they know how to talk. Maybe they'll get some extra gum for the kids, or a box of clothes. And this is right next to a deep philosophy—their relationship with death, with time. It's not for some gum and German chocolate that they refuse to leave these peasant huts they've been living in their whole lives.

On the way back, the sun is setting, I say, "Look at how beautiful this land is!" The sun is illuminating the forest and the fields, bidding us farewell. "Yes," one of the Germans who speaks

Russian answers, "it's pretty, but it's contaminated." He has a dosimeter in his hand. And then I understand that the sunset is only for me. This is my land. I'm the one who lives here.

*Natalya Arsenyevna Roslova, head of the Mogilev Women's Committee for the Children of Chernobyl*

CHILDREN'S CHORUS

*Alyosha Belskiy, 9; Anya Bogush, 10; Natasha Dvoretskaya, 16; Lena Zhudro, 15; Yura Zhuk, 15; Olya Zvonak, 10; Snezhana Zinevich, 16; Ira Kudryacheva, 14; Ylya Kasko, 11; Vanya Kovarov, 12; Vadim Karsnosolnyshko, 9; Vasya Mikulich, 15; Anton Nashivankin, 14; Marat Tatartsev, 16; Yulia Taraskina, 15; Katya Shevchuk, 15; Boris Shkirmankov, 16.*

There was a black cloud, and hard rain. The puddles were yellow and green, like someone had poured paint into them. They said it was dust from the flowers. Grandma made us stay in the cellar. She got down on her knees and prayed. And she taught us, too. "Pray! It's the end of the world. It's God's punishment for our sins." My brother was eight and I was six. We started remembering our sins. He broke the glass can with the raspberry jam, and I didn't tell my mom that I'd got my new dress caught on a fence and it ripped. I hid it in the closet.

\*

Soldiers came for us in cars. I thought the war had started. They were saying these things: "deactivation," "isotopes." One soldier was chasing after a cat. The dosimeter was working on the cat like an automatic: click, click. A boy and a girl were chasing the cat, too. The boy was all right, but the girl kept crying, "I won't give him up!" She was yelling: "Run away, run little girl!" But the soldier had a big plastic bag.

*

I heard—the adults were talking—Grandma was crying—since the year I was born [1986], there haven't been any boys or girls born in our village. I'm the only one. The doctors said I couldn't be born. But my mom ran away from the hospital and hid at Grandma's. So I was born at Grandma's. I heard them talking about it.

I don't have a brother or sister. I want one.

Tell me, lady, how could it be that I wouldn't be born? Where would I be? High in the sky? On another planet?

*

The sparrows disappeared from our town in the first year after the accident. They were lying around everywhere—in the yards, on the asphalt. They'd be raked up and taken away in the containers with the leaves. They didn't let people burn the leaves that year, because they were radioactive, so they buried the leaves.

The sparrows came back two years later. We were so happy, we were calling to each other: "I saw a sparrow yesterday! They're back."

The May bugs also disappeared, and they haven't come back. Maybe they'll come back in a hundred years or a thousand. That's what our teacher says. I won't see them.

*

September first, the first day of school, and there wasn't a single flower. The flowers were radioactive. Before the beginning of the year, the people working weren't masons, like before, but soldiers. They mowed the flowers, took off the earth and took it away somewhere in cars with trailers.

In a year they evacuated all of us and buried the village. My father's a cab driver, he drove there and told us about it. First they'd tear a big pit in the ground, five meters deep. Then the firemen would come up and use their hoses to wash the house from its roof to its foundation, so that no radioactive dust gets kicked up. They wash the windows, the roof, the door, all of it. Then a crane drags the house from its spot and puts it down into the pit. There's dolls and books and cans all scattered around. The excavator picks them up. Then it covers everything with sand and clay, leveling it. And then instead of a village, you have an empty field. They sowed our land with corn. Our house is lying there, and our school and our village council office. My plants are there and two albums of stamps, I was hoping to bring them with me. Also I had a bike.

\*

I'm twelve years old and I'm an invalid. The mailman brings two pension checks to our house—for me and my granddad. When the girls in my class found out that I had cancer of the blood, they were afraid to sit next to me. They didn't want to touch me.

The doctors said that I got sick because my father worked at Chernobyl. And after that I was born. I love my father.

\*

They came for my father at night. I didn't hear how he got packed, I was asleep. In the morning I saw my mother was crying. She said, "Papa's in Chernobyl now."

We waited for him like he was at the war.

He came back and started going to the factory again. He didn't tell us anything. At school I bragged to everyone that my father just came back from Chernobyl, that he was a liquidator, and the liquidators were the ones who helped clean up after the accident. They were heroes. All the boys were jealous.

A year later he got sick.

We walked around in the hospital courtyard—this was after his second operation—and that was the first time he told me about Chernobyl.

They worked pretty close to the reactor. It was quiet and peaceful and pretty, he said. And as they're working, things are happening. The gardens are blooming. For who? The people have left the villages. They "cleaned" the things that needed to be left behind. They took off the topsoil that had been contaminated by cesium and strontium, and they washed the roofs. The next day everything would be "clicking" on the dosimeters again.

"In parting they shook our hands and gave us certificates of gratitude for our self-sacrifice." He talked and talked. The last time he came back from the hospital, he said: "If I stay alive, no more physics or chemistry for me. I'll leave the factory. I'll become a shepherd." My mom and I are alone now. I won't go to the technical institute, even though she wants me to. That's where my dad went.

*

I used to write poems. I was in love with a girl. In fifth grade. In seventh grade I found out about death.

I read in Garcia Lorca: "the cry's black root." I began to learn how to fly. I don't like playing that game, but what can you do?

I had a friend, Andrei. They did two operations on him and then sent him home. Six months later he was supposed to get a third operation. He hanged himself from his belt, in an empty classroom, when everyone else had gone to gym glass. The doctors had said he wasn't allowed to run or jump.

Yulia, Katya, Vadim, Oksana, Oleg, and now Andrei. "We'll die, and then we'll become science," Andrei used to say. "We'll die and everyone will forget us," Katya said. "When I die, don't bury me at the cemetery, I'm afraid of the cemetery, there are only dead people and crows there," said Oksana. "Bury me in the field." Yulia used to just cry. The whole sky is alive for me now when I look at it, because they're all there.

## A SOLITARY HUMAN VOICE

Not long ago I was so happy. Why? I've forgotten. It feels like another life now. I don't even understand, I don't know how I've been able to begin living again. Wanting to live. But here I am. I laugh, I talk. I was so heartbroken, I was paralyzed. I wanted to talk with someone, but not anyone human. I'd go to a church, it's so quiet there, like in the hills. So quiet, you can forget your life there. But then I'd wake up in the morning, my hand would feel around—where is he? It's his pillow, his smell. There's a tiny bird running around on the windowsill making the little bell ring, and it's waking me up, I've never heard that sound before, that voice. Where is he? I can't tell about all of it, I can't talk about all of it. I don't even understand how I stayed alive. In the evening my daughter would come up to me: "Mom, I'm already done with my homework." That's when I remember I have kids. But where is *he*? "Mom, my button fell off. Can you sew it back?" How do I go after him, meet up with him? I close my eyes and think of him until I fall asleep. He comes to me in my sleep, but only in flashes, quickly. Right away he disappears. I can even hear his footsteps. But where does he go? Where? He didn't want to die. He looks out the window and looks, looks at the sky. I put one pillow under him, then another, then a third. So that he could be high up. He died for a long time. A whole year. We couldn't part. [*She is silent for a long time.*]

No, don't worry, I don't cry anymore. I want to talk. I can't tell myself that I don't remember anything, the way others do. Like my friend. Our husbands died the same year, they were in Chernobyl together, but she's already planning to get married. I'm not condemning her—that's life. You need to survive. She has kids.

He left for Chernobyl on my birthday. We had guests over, at the table, he apologized to them. He kissed me. But there was already a car waiting for him outside the window.

It was October 19, 1986, my birthday. He was a construction worker, he traveled all over the Soviet Union, and I waited for him. That's just how we lived over the years—like lovebirds. We'd say goodbye and then we'd reunite. And then—this fear came over our mothers, his and mine, but we didn't feel it. Now I wonder why. We knew where he was going. I could have taken the neighbor boy's tenth-grade physics textbook and taken a look. He didn't even wear a hat. The rest of the guys he went with lost their hair a year later, but his grew out really thick instead, like a mane. None of those boys is alive anymore. His whole brigade, seven men, they're all dead. They were young. One after the other. The first one died after three years. We thought: well, a coincidence. Fate. But then the second died and the third and the fourth. Then the others started waiting for their turn. That's how they lived. My husband died last. He worked high in the air. They'd turn the lights off in evacuated villages and climb on the light poles, over the dead houses, the dead streets, always high up in the air. He was almost two meters tall, he weighed ninety kilograms—who could kill him? [*Suddenly, she smiles.*]

Oh, how happy I was! He came back. We had a party, whenever he came back there was a party. I have a nightgown that's so long, and so beautiful, I wore it. I liked expensive lingerie, everything I have is nice, but this nightgown was special—it was for special occasions. For our first day, our first night. I knew his whole body by heart, all of it, I kissed all of it. Sometimes I'd dream that I was part of his body, that we're inseparable that way. When he was gone I'd miss him so much, it would be physically painful. When we parted, for a while I'd feel lost, I wouldn't know what street I was on, what the time was.

When he came back he had knots in his lymph nodes, they were small but I heard them with my lips. "Will you go to a doctor?" I said. He calmed me down: "They'll go away." "What was it like in Chernobyl?" "Just ordinary work." No bravado, no panic. I got one thing out of him: "It's the same there as it is here." In the cafeteria there they'd serve the ordinary workers with noodles and canned foods on the first floor, and then the bosses and generals would be served fruit, red wine, mineral water on the second. Up there they had clean tablecloths, and a dosimeter for every man. Whereas the ordinary workers didn't get a single dosimeter for the whole brigade.

Oh, how happy I was! We still went to the sea then, and the sea was like the sky, it was everywhere. My friend went with her husband, too, and she thinks the sea was dirty—"We were afraid we'd get cholera." It's true, there was something about that in the papers. But I remember it differently, in much brighter lights. I remember that the sea was everywhere, like the sky. It was blue-blue. And he was nearby.

I was born for love. In school all the girls dreamt of going to the university, or on a Komsomol work trip, but I dreamt of getting married. I wanted to love, to love so strongly, like Natasha Rostov. Just to love. But I couldn't tell anyone about it, because back then you were only supposed to dream of the Komsomol construction trip. That's what they taught us. People were rearing to go to Siberia, to the impenetrable taiga, remember they'd sing: "past the fog and the smell of the taiga." I didn't get into the university during the first year, I didn't have enough points at the exams, and I went to work at the communications station. That's where I met him. And I proposed to him myself, I asked him: "Marry me. I love you so much!" I fell in love up to my ears. He was such a great-looking guy. I was flying through the air. I asked him myself: "Marry me." [*Smiles.*]

Another time I'll think about it and find ways of cheering myself up—like, maybe death isn't the end, and he's only changed somehow and lives in another world. I work in a library, I read a lot of books, I meet many people. I want to talk about death, to understand it. I'm looking for consolation. I read in the papers, in books, I go to the theater if it's about death. It's physically painful for me to be without him—I can't be alone.

He didn't want to go to the doctor. "I don't hear anything, and it doesn't hurt." But the lymph nodes were already the size of eggs. I forced him into the car and took him to the clinic. They referred him to an oncologist. One doctor looked at him, called over another: "We have another Chernobylite here." After that they didn't let him go. A week later they did an operation: they took out the thyroid gland and larynx and replaced them with some tubes. Yes . . . [*She is silent.*] Yes—now I know that

that, too, was a happy time. God! The errands I sent myself on—running to the stores buying presents for the doctors, boxes of chocolates, imported liqueurs. I got chocolates for the nurses. And they took them. Meanwhile he's laughing at me: "Understand, they're not gods. And they've got enough chemo and radiation to go around. They'll give me that without the candies." But I was running to the other side of town for some soufflé cake or French perfume. In those times you couldn't get that stuff without knowing someone, it was all under the counter. That was right before they sent him home.

They sent him home! They gave me a special needle and showed me how to use it. I was supposed to feed him through the needles. I learned how to do everything. Four times a day I'd cook up something fresh, it had to be fresh, and I'd grind it all up in the meat grinder, put it onto a stringer, and then pour this all into the needle. I'd stick the needle into the biggest of his tubes, and that one went into his stomach. But he'd lost his sense of smell by then. I'd say, "How does this taste?" He wouldn't know.

We still went to the movies a few times. And we'd kiss there. We were hanging by such a thin thread, but we thought we were digging into life again. We tried not to talk about Chernobyl, not to remember it. It was a forbidden topic. I wouldn't let him come to the phone, I'd intercept it, all his boys were dying one by one. It was a forbidden topic. And then one morning I wake him up and give him his robe, but he can't get up. And he can't talk. He's stopped talking. His eyes are huge. That's when he got scared. Yes . . . [*She is silent again.*]

We had one year left after that. He spent that whole year dying. He got worse with each day, but he didn't know that his boys were dying, too. That's what we lived with—with that thought. It's impossible to live with, too, because you don't know what it is. They say, "Chernobyl," and they write, "Chernobyl." But no one knows what it is. Something frightening opened up before us. Everything is different for us: we aren't born the same, we don't die the same. If you ask me, How do people die after Chernobyl? The person I loved more than anything, loved him so much that I couldn't possibly have loved him more if I'd given birth to him myself—turned—before my eyes—into a monster. They'd taken out his lymph nodes, so they were gone and his circulation was disrupted, and then his nose kind of shifted, it grew three times bigger, and his eyes became different—they sort of drifted away, in different directions, there was a different light in them now, and I saw expressions in them I hadn't seen, as if he was no longer himself but there was still someone in there looking out. Then one of the eyes closed completely.

And what was I afraid of? Only that he'd see it. But he began asking me, showing me with his hands, like, bring me the mirror. I'd run off to the kitchen, as if I'd forgotten, or I hadn't heard, or something else. I did that for two days, tricking him, but on the third he wrote in his notebook in large letters with three exclamation points: "Bring the mirror!!!" We already had a notebook, a pen, a pencil, that's how we communicated because by then he couldn't even whisper. He was completely mute. I ran off to the kitchen, started banging on the pots and pans. As if I hadn't read what he'd written, or I'd misunderstood. He writes again: "Bring the mirror!!!" With those exclamation points.

So I brought him the mirror, the smallest one. He looked, and then he grabbed his head and he rocked, rocked back and forth on the bed. I start pleading with him—"As soon as you get a little better, we'll go off to a village together, an abandoned village. We'll buy a house and we'll live there, if you don't want to live in a big town with a lot of people. We'll live by ourselves." And I wasn't kidding, I would have gone with him anywhere, just so he'd be there, anywhere. He was all that mattered. I wasn't kidding.

I won't remember anything I wouldn't want to talk about. But everything happened. I looked very far, maybe further than death. [*Stops.*]

I was sixteen when we met, he was older than me by seven. We dated for two years. I love that neighborhood in Minsk near the main post office, Volodarkovo Street, we'd meet there under the big clock. I lived near the worst industrial complex and I'd take bus #5, which didn't stop near the main post office but went a little further, to the children's clothing store. I'd always be just a tiny bit late so I could go by on the bus and see him there and think: oh, what a handsome man is waiting for me! I didn't notice anything for two years, not the winter, not the summer. He took me to concerts, to Edith Piekha, my favorite. We didn't go dancing, he didn't know how. We kissed, we just kissed. He called me "my little one." My birthday, it was always my birthday, it's strange but the most important things happened to me on my birthday, and after that try not believing in fate. I was waiting for him under the clock, we were supposed to meet at five but he wasn't there. At six I'm upset, I wander over to my stop in tears, cross the street, then I look around because I feel something—and there he is running after me, against the light, in his work clothes, in his boots. That's how I liked him best, in his hunter's jacket, a sailor's shirt—everything looked good on him. We went to his house, he changed and we decided to celebrate my birthday in a restaurant. But we couldn't get in, it was evening and slipping the maitre d' a five or a ten, like other people, neither of us knew how to do that. "Okay," he said, he was shining, "let's buy some champagne and some cake and go to the park, that's where we'll celebrate." Under the stars, under the sky! That's how he was. We sat on a bench in Gorky Park until morning. I never had another birthday like that, and that's when I said to him: "Marry me. I love you so much!" He laughed, "But you're still little." But the next day we went to register.

I was so happy! I wouldn't change anything in my life, even if someone had told me, from above . . . On the day we got married he couldn't find his passport, we turned the whole house over. They wrote us down at the registry on a piece of paper, temporarily. "My daughter, it's a bad sign," my mom cried. We found the passport later in some old pants of his in the attic. Love! It wasn't even love, it was a long falling in love. I used to dance in front of the mirror in the mornings—I'm young, I'm pretty, he loves me! Now I forget that face—the face I had then, with him. I don't see that face anymore in the mirror.

Is this something I can talk about? Call it with words? There are secrets—I still don't understand what that was. Even in our last month, he'd still call for me at night. He felt desire. He loved me more than he did before. During the day, I'd look at him, and I couldn't believe what had happened at night. We didn't want to part. I caressed him, I petted him. In those minutes I remembered the happiest times, the light. Like when he came back from Kamchatka with a beard, he'd grown a beard there. My birthday in the park on the bench. "Marry me." Do I need to talk about it? Can I? I myself went to him the way a man goes to a woman. What could I give him aside from medicine? What hope? He didn't want to die.

But I didn't tell my mother anything. She wouldn't have understood me. She would have judged me, cursed us. Because this wasn't just an ordinary cancer, which everyone already is afraid of, but Chernobyl cancer, even worse. The doctors told me: if the tumors had metastasized within his body, he'd have died quickly, but instead they crawled upward, along the body, to the face. Something black grew on him. His chin went somewhere, his neck disappeared, his tongue fell out. His veins popped, he began to bleed. From his neck, his cheeks, his ears. To all sides. I'd bring cold water, put wet rags against him, nothing helped. It was something awful, the whole pillow would be covered in it. I'd bring a washbowl from the bathroom, and the streams would hit it, like into a milk pail. That sound, it was so peaceful and rural. Even now I hear it at night. While he was still conscious, if he started clapping, that was our sign: Call the ambulance. He didn't want to die. He was forty-five years old. I'd call the ambulance, and they know us, they don't want to come: "There's nothing we can do for your husband." Just give him a shot! Some narcotic. I learned how to do it myself, but the shot leaves a bruise under the skin, and it doesn't go away.

One time I managed to get an ambulance, it comes with a young doctor. He comes over and right away staggers back. "Excuse me, he's not from Chernobyl is he? One of the ones who went there?" I say, "Yes." And he, I'm not exaggerating, he cries out: "Oh, dear woman, then let this end quickly! Quickly! I've seen how the ones from Chernobyl die." Meanwhile my husband is conscious, he hears this. At least he doesn't know, he hasn't guessed, that he's the last one from his brigade still alive.

Another time the nurse from the nearby clinic comes, she just stands in the hallway and refuses to come in. "Oh, I can't!" she says. And I can? I can do anything. What can I think of? How can I save him? He's yelling, he's in pain, all day he's yelling. Finally I found a way: I filled a syringe with vodka and put that in him. He'd turn off, forget the pain. I didn't think of it myself, some other women told me, they'd been through the same thing.

His mom used to come: "Why'd you let him go to Chernobyl? How could you?" It didn't even occur to me then that I could keep him from going, and for him, he probably didn't think it was possible to refuse. That was a different time, a military time. I asked him once: "Are you sorry now that you went there?" He shakes his head no. He writes in his notebook: "When I die, sell the car and the spare tire, and don't marry Tolik." That was his brother, Tolik. He liked me.

There are things—I'm sitting next to him, he's asleep, and he has such pretty hair. I took some scissors and quietly cut off a lock of it. He opened his eyes and looked and saw what I had in my hand, he smiled. I still have his watch, his military ID, and his medal from Chernobyl. [*She is silent.*] I was so happy! I'd spend hours in the maternity ward, I remember, sitting at the window, waiting for him and looking out. I didn't really understand what was happening: What's wrong with me? I couldn't get enough of looking at him. I figured it would end at some point. I'd make him some food in the morning and then wonder at him eating it. And him shaving, him walking down the street. I'm a good librarian, but I don't understand how you can love your job. I loved only him. Him alone. And I can't be without him. I yell at night. I yell into the pillow, so that my kids don't hear.

It never occurred to me that he'd leave the house, that we'd be apart for the end. My mom, his brother, they started telling me, hinting, that the doctors, like, were advising, you know, in short, there was a place near Minsk, a special hospital, where hopeless cases like this used to die—they used to be the soldiers from Afghanistan, without arms, without legs—now they sent the ones from Chernobyl there. They're begging me: it'll be better there, there are always doctors around. I didn't want to, I didn't even want to hear about it. Then they convinced him, and he started asking me. "Take me there. Stop torturing your-self." Meanwhile I'm asking for sick leave, or just for personal leave, on my own account, because you're only supposed to take sick leave in order to care for a sick child, and personal leave isn't supposed to be longer than a month. But he wrote all over his notebook, made me promise that I'd take him there. Finally I go in a car with his brother. It's at the edge of a village, it's called Grebenka, a big wooden house, a broken well next to it. The toilet is outside. Some old ladies in black—religious women. I didn't even get out of the car. I didn't get up. That night I kissed him: "How could you ask me for that? I'll never let it happen! Never!" I kissed him all over.

The last few weeks were the scariest. We spent half an hour peeing into a half-liter can. He kept his eyes down the whole time, he was ashamed. "Oh, how can you think like that?!" I said. I kissed him. On the last day, there was this moment, he opened his eyes, sat up, smiled and said: "Valyushka!" That's me. He died alone. As everyone dies alone. They called me from work. "We'll bring his red certificate of achievement." I ask him: "Your boys want to come, they want to bring your certificate." He shakes his head: no. But they came by anyway. They brought some money, and the certificate in a red folder with a photo of Lenin on the front. I took it and thought: "So is this what he's dying for? The papers are saying that it's not just Chernobyl, all of Communism is blowing up. But the picture is the same." The guys wanted to say something nice to him, but he just covered himself with the blanket so that only his hair was sticking out. They stood there awhile and then they left. He was already afraid of people. I was the only one he wasn't afraid of. When we buried him, I covered his face with two handkerchiefs. If someone asked me to, I lifted them up. One woman fainted. And she used to be in love with him, I

was jealous of her once. "Let me look at him one last time."
"All right." I didn't tell her that when he died no one wanted to
come near him, everyone was afraid. According to our customs
you're not supposed to wash and clothe your relatives. Two
orderlies came from the morgue and asked for vodka. "We've
seen everything," they told me, "people who've been smashed
up, cut up, the corpses of children caught in fires. But nothing
like this. The way the Chernobylites die is the most frightening
of all." [*Quietly.*] He died and he lay there, he was so hot. You
couldn't touch him. I stopped the clocks in the house when he
died. It was seven in the morning.

Those first days without him, I slept for two days straight,
no one could wake me, I'd get up, drink some water, not eat
anything, and fall back down on the pillow. Now I find it
odd, inexplicable: how could I go to sleep? When my friend's
husband was dying, he threw dishes at her: why was she so
pretty and young? But mine just looked at me and looked. He
wrote down in his notebook: "When I die, burn the remains.
I don't want you to be afraid." There'd been rumors that even
after dying the men from Chernobyl are radioactive. I read
that the graves of the Chernobyl firefighters who died in the
Moscow hospitals and were buried near Moscow at Mitino
are still considered radioactive, people walk around them and
don't bury their relatives nearby. Even the dead fear these dead.
Because no one knows what Chernobyl is. People have guesses
and feelings. He brought back the white costume he worked in
from Chernobyl. The pants, the special protection. And that
suit stayed up on our storage space right up until he died. Then
my mother decided, "We need to throw out all his things." She
was afraid. But I wanted to keep even that suit. Which was
criminal—I had children at home. So we took all the things
outside of town and buried them. I've read a lot of books, I live

among books, but nothing can explain this. They brought me the urn. I wasn't afraid, I felt around with my hand, and there was something tiny, like seashells in the sand, those were his hip bones. Before that I'd touched his things but never heard him or felt him, but here I did. I remember the night, after he died, I sat next to him—and suddenly I saw this little puff of smoke—I saw it again at the crematorium—it was his soul. No one saw it except me. And I felt like we'd seen each other one more time.

I was so happy! He'd be off on a work trip, I'd count the days and hours until he came back. I, physically, can't be without him. We'd go visit his sister in the country, in the evening she'd say, "I put the bedding down for you in that room, and for you in the other room." We'd look at each other and laugh—we couldn't imagine sleeping in different rooms. I can't be without him. We eloped. His brother too. They're so alike. But if anyone touched me now, I think, I'd cry and cry.

Who took him away from me? By what right? They brought a notice with a red banner across the top on October 19, 1986, like they were calling him up for the war.

[*We drink tea and she shows me the family photographs, the wedding photographs. And then, as I'm getting up to go, she stops me.*]

How am I going to live now? I haven't told you everything, not to the end. I was very happy, insanely happy. Maybe you shouldn't write my name? There are secrets—people pray in secret, in a whisper just to themselves. [*Stops.*] No, write down my name, let God know. I want to understand. And I want to understand why we need to suffer. What for? At first I thought that I'd have something new in my vision, something dark and not-mine. What saved me? What pushed me back out into life? My son did. I have another son, our son, he's been sick for a long time. He grew up but he sees the world with the eyes of a child, a five-year-old. I want to be with him. I'm hoping to trade my apartment for one closer to Novinki, that's where we have the mental hospital. He's there. The doctors ordered it: if he's to live, he needs to live there. I go there on the weekends. He greets me: "Where's Papa Misha? When will he come?" Who else is going to ask me about that? He's waiting for him.

We'll wait for him together. I'll read my Chernobyl prayer in a whisper. You see, he looks at the world with the eyes of a child.

*Valentina Timofeevna Panasevich,*
*wife of a liquidator*

# IN PLACE OF AN EPILOGUE

I used to travel among other people's suffering, but here I'm just as much a witness as the others. My life is part of this event. I live here, with all of this.

There are 350 atomic bombs in our land. People are already living after the nuclear war—though when it began, they didn't notice.

Now people come here from other wars. Thousands of Russian refugees from Armenia, Georgia, Abkhazia, Tajikistan, Chechnya—from anywhere where there's shooting, they come to this abandoned land and the abandoned houses that weren't destroyed and buried by special squadrons. There are over 25 million ethnic Russians outside of Russia—a whole country—and there's nowhere for some of them to go but Chernobyl. All the talk about how the land, the water, the air can kill them sounds like a fairy tale to them. They have their own tale, which is a very old one, and they believe in it—it's about how people kill one another with guns.

I used to think I could understand everything and express everything. Or almost everything. I remember when I was writing my book about the war in Afghanistan, *Zinky Boys*, I went to Afghanistan and they showed me some of the foreign weapons that had been captured from the Afghan fighters. I was amazed at how perfect their forms were, how perfectly a human thought had been expressed. There was an officer standing next to me and he said, "If someone were to step on this Italian mine that you say is so pretty it looks like a Christmas decoration, there would be nothing left of them but a bucket of meat. You'd have to scrape them off the ground with a spoon." When I sat down to write this, it was the first time I thought, "Is this something I should say?" I had been raised on great Russian literature, I thought you could go very very far, and so I wrote about that meat. But the Zone—it's a separate world, a world within the rest of the world—and it's more powerful than anything literature has to say.

For three years I rode around and asked people: the workers at the nuclear plant, the scientists, the former Party bureaucrats, doctors, soldiers, helicopter pilots, miners, refugees, re-settlers. They all had different fates and professions and temperaments. But Chernobyl was the main content of their world. They were ordinary people answering the most important questions.

I often thought that the simple fact, the mechanical fact, is no closer to the truth than a vague feeling, rumor, vision. Why repeat the facts—they cover up our feelings. The development of these feelings, the spilling of these feelings past the facts, is what fascinates me. I try to find them, collect them, protect them.

These people had already seen what for everyone else is still unknown. I felt like I was recording the future.

*Svetlana Alexievich*

## LANNAN SELECTIONS

The Lannan Foundation, located in Santa Fe, New Mexico, is a family foundation whose funding focuses on special cultural projects and ideas which promote and protect cultural freedom, diversity, and creativity.

The literary aspect of Lannan's cultural program supports the creation and presentation of exceptional English-language literature and develops a wider audience for poetry, fiction, and nonfiction.

Since 1990, the Lannan Foundation has supported Dalkey Archive Press projects in a variety of ways, including monetary support for authors, audience development programs, and direct funding for the publication of the Press's books.

In the year 2000, the Lannan Selections Series was established to promote both organizations' commitment to the highest expressions of literary creativity. The Foundation supports the publication of this series of books each year, and works closely with the Press to ensure that these books will reach as many readers as possible and achieve a permanent place in literature. Authors whose works have been published as Lannan Selections include Ishmael Reed, Stanley Elkin, Ann Quin, Nicholas Mosley, William Eastlake, and David Antin, among others.

# SELECTED DALKEY ARCHIVE PAPERBACKS

## FOR A FULL LIST OF PUBLICATIONS, VISIT:
## www.dalkeyarchive.com

# SELECTED DALKEY ARCHIVE PAPERBACKS

## FOR A FULL LIST OF PUBLICATIONS, VISIT:
## www.dalkeyarchive.com